Rubber Plantation in Tripura
Changing Livelihood

THE AUTHORS

Dr. Raju Debbarma has completed his M.A in Geography, M.Phil and Ph.D. from the Department of Geography, North-Eastern Hill University, Shillong. He has participated in national and international seminars and workshops. He has published research papers in National and International journals and in edited volumes. His interest area of research includes rubber plantation and its related issues in the state of Tripura.

Dr. (Mrs.) Subrata Purkayastha (Ph.D.), an ardent researcher, teaches Geography in North-Eastern Hill University, Shillong one of the premier institute of North-East India. She specializes in population and cultural geography having a keen interest on spatio-temporal changes brought about by development activities in North-East of India and the response of especially the various smaller communities who till recently were living in close proximity to nature deriving their livelihood from direct contact with natural resources. Dr. Purkayastha has published numerous research papers in leading professional journals and jointly edited a book titled -Rivers and riverine landscape in North-East India.

Rubber Plantation in Tripura
Changing Livelihood

Raju Debbarma
Subrata Purkayastha

2022
Regency Publications
A Division of
Astral International Pvt. Ltd.
New Delhi – 110 002

ISBN: 9789354616501 (HB)

Published by : **Regency Publications**
A Division of
Astral International Pvt. Ltd.
– ISO 9001:2015 Certified Company –
4736/23, Ansari Road, Darya Ganj
New Delhi-110 002
Ph. 011-43549197, 23278134
E-mail: info@astralint.com
Website: www.astralint.com

Prasad Rao Vaddarapu, IFS
Managing Director, TRPC Ltd.
A PSU Under Govt. of Tripura , Agartala

Foreword

Rubber plantation in Tripura was initiated in 1963; since then, development of rubber plantation was extensively distributed by the Rubber Board, Tripura Forest Development and Plantation Corporation Limited (TFDPC Ltd.), Tripura Rehabilitation Plantation Corporation Limited (TRPC Ltd.) and others departments with the ardent support given by the Govt. of Tripura. Tripura now has the 2nd largest area under rubber plantation and production after Kerala state in India. The plantation has provided tremendous opportunity of employment and livelihoods in the state, improving the social and economic status of the many beneficiaries and growers in the rural areas of the state. The book "Rubber Plantation in Tripura: Changing Livelihood" prepared by Dr. Raju Debbarma and Dr. Subrata Purkayastha is specially about the rubber plantation and its impact on the socio-economic development of the beneficiaries under the TRPC Ltd. in the state of Tripura. I sincerely appreciate Dr. Raju Debbarma and Dr. Subrata Purkayastha for their efforts to work on such resourceful thesis and deserve commendation. I am sure that this book would be great helpful to scholars, academicians and readers in providing information about the impacts and contributions of rubber plantation to the people of the state of Tripura.

Agartala

Date: 19.03.2021

Managing Director
TRPC Ltd., Govt. of Tripura, Agartala

Acknowledgment

The present book is an outcome of an in-depth research of us, in assessing the role of rubber plantation scheme for the socio-economic upliftment of the marginalized population of Tripura. We would like to acknowledge Sri Prasada Rao Vaddarapu Managing Director TRPC limited Agartala, for providing valuable information without which this book would not have seen the light of the day. We would also fail in our duty if we do not acknowledge the villagers of Depachhara, Dakshin Dhumachhara, Pathalighat and Dakshin Taidu where field work were conducted and empirical data generated to assess the impact of rubber plantation scheme at grass root level. Above all our profound thanks to Astral International Pvt. Ltd. New Delhi, for publishing this book.

Raju Debbatma
Subrata Purkayastha

Preface

This book addresses the role of different agencies in the expansion of area under rubber plantation in the state of Tripura, the second largest producer of natural rubber in India. It presents important facts related to diverse livelihood opportunities of the households rehabilitated through the rubber plantation scheme. In 1960s Rubber trees were introduced in Tripura, a non-traditional rubber growing state, to help in the rejuvenation of the forests impacted by jhum cultivation and rehabilitate the landless jhumia families into settled agriculture. With time the economic and ecological benefits of the rubber trees attracted the attention of both the private and public sectors. This resulted in the commercial plantation of rubber trees both by the individual/private owners and public sector undertakings. Various Public Agencies emerged, that initiated Rubber plantation-based rehabilitation scheme to improve the socio-economic status of the marginalised landless population including the jhumias, who were lured into settled agriculture. The book critically assess how far the Rubber plantation scheme has enabled the jhumia families to become the owner of the rubber estates.

Furthermore, it provides an insight on the rehabilitation scheme through Rubber Plantation, from its inception to its success story in the state of Tripura. The growth and emerging pattern of distribution of rubber plantation, leading to concentration of area under rubber plantation in Tripura has been assessed by using satellite data through RS and GIS technique. The role and contribution of different agencies in the growth and distribution of the area under rubber plantation has also been measured. The impact of rubber plantation on the livelihood practices, livelihood diversification has been evaluated. The five livelihood assets/capitals of the households practicing rubber plantation has been measured in order to ascertain the level of socio-economic changes brought about by the Rubber plantation Scheme. The book is content-rich based on both secondary and empirical data, that has been collected at village and household levels. The authors are of the view that the book will give a holistic insight into the growth and distribution of Rubber plantation and associated changes in livelihood practices of the people in Tripura.

Dr. Raju Debbarma

Dr.(Mrs.) Subrata Purkayastha

Contents

	Foreword	*v*
	Acknowledgment	*vii*
	Preface	*ix*
1.	**Introduction**	**1**
	Functioning of Rubber Plantation Scheme, History of Rubber Plantation in the World, Global Level, National and regional level, Local Level, Focus area of study addressed in the present book, Database and Method Used in the Study, Source of Secondary Data, Method used for Analyzing Secondary Data, Image Pre-processing, Image Processing, Image Classification, Accuracy Assessment, Rubber Plantation Map, Source of Primary Data, Methodology for Generating Primary Data, Selection of District, Selection of Agency for the Study, Selection of Field Study Site, Selection of Households and Sampling Technique, Sampling size, Data Collection, Concept of Sustainable Livelihood, Analysis of Primary Data, Livelihood Diversification, Five capitals of the livelihood assets, Limitations, Some of the limitations are, Design of the Book	
2.	**Geographical Background of Tripura**	**49**
	Introduction, Brief History, Geographical setting, Location and Area, Physiography, Geology, Climate, Rivers, Soil, Vegetation, Mineral resources, Population, Tribal composition, Economy, Socio-economic scenario	
3.	**Growth and Distribution of Rubber Plantation in Tripura**	**67**
	Introduction, Rubber plantation in India (Colonial phase), Rubber plantation after Indian Independence, Rubber plantation in Tripura, Rubber Plantation as Rehabilitation Programme, Emerging pattern of increase of area under rubber plantation in Tripura, Share of area under rubber plantation of Tripura to the total area under rubber plantation of the country, Expansion of area under rubber plantation in Tripura, Nature of growth of area under rubber plantation in Tripura, Decadal growth of area under rubber plantation in Tripura, Annual gain in area under rubber plantation in Tripura, Area under rubber plantation and production in Tripura, Distribution of area under rubber plantation in Tripura, District wise growth rate of area under rubber plantation in Tripura, District wise concentration of area under rubber plantation in Tripura, Spatial distribution of rubber plantation in Tripura, Conclusion	

4. **Role of Different Agencies on the Growth and Distribution of Rubber Plantation in Tripura** **107**

Introduction, Role of different rubber growing agencies in Tripura, Tripura Forest Development and Plantation Corporation Limited (TFDPCL), Aims of the TFDPCL, Functioning of the TFDPCL, Growth of rubber plantation under the TFDPCL, Role of TFDPCL in the expansion of area under rubber plantation, Role of TFDPCL in the annual gain of area under rubber plantation, Distribution of area under rubber plantation developed by the TFDPCL, District wise concentration of area under rubber plantation developed by the TFDPCL (2014-15), Level of achievement of the basic objectives of the TFDPCL, Tripura Rehabilitation Plantation Corporation Limited (TRPCL), Aims of the TRPCL, Functioning of the TRPCL, TRPCL and the emerging pattern of increase of area under rubber plantation, Share of area under rubber plantation developed by the TRPCL to the total area under rubber plantation of the state, Role of TRPCL in the expansion of area under rubber plantation, Role of TRPCL in the annual gain in area under rubber plantation, Distribution of area under rubber plantation developed by the TRPCL, District wise concentration of area under rubber plantation developed by the TRPCL,The Rubber Board Functioning of the Rubber Board, Rehabilitation scheme of the Rubber Board, Functioning of the project of the Rubber Board, Rubber Board and the emerging pattern of increase of area under rubber plantation , Share of area under rubber plantation developed by the Rubber Board to the total area under rubber plantation of the state, Role of Rubber Board in the expansion of area under rubber plantation, Role of Rubber Board in the annual gain of area under rubber plantation, Distribution of area under rubber plantation developed by the Rubber Board, District wise concentration of area under rubber plantation developed by the Rubber Board, Role of other agencies, Comparison among the different agencies in the expansion of area under rubber plantation in Tripura, Conclusion

5. **Impact of Rubber Plantation on the Socio-Economic Conditions and Livelihoods of the Surveyed Households** **149**

Introduction, RPC/village wise tribe composition and the livelihood assets/ capitals of the BHHs, Tribe composition of the BHHs, Comparative analysis of the five livelihood assets/capitals of the BHHs living in the four surveyed RPCs/ villages, Human asset/capital, Social asset/capital, Natural asset/capital, Physical asset/capital, Financial asset/capital, Socio-economic characteristics of the BHHs and NBHHs, Tribe composition, Head of the households, Size of the households, Literacy and levels of education, Access to amenities, house type and ownership of consumable commodities, Household income, Major livelihood practices of the BHHs and NBHHs, Livelihood practices, Livelihood diversification, Comparative analysis of the five livelihood assets of the BHHs and NBHHs, Comparative analysis of the livelihood assets/capitals of BHHs and NBHHs

6. **Summary of Findings and Conclusions** **183**

Bibliography **192**

Appendices **203**

Color Plate **231**

1
Introduction

The hilly state of Tripura located in North east India is of strategic importance as 84% of its border is international, being surrounded by Bangladesh on its north, south and west, the state is connected with the Indian union through a narrow-forested strip passing through the Cachar district of Assam. The partition of India and Pakistan in 1947, changed the boundary of erstwhile Princely state of Tripura. that had a larger boundary stretching over the adjoining plains of present Bangladesh. Tripura, an integral part of undivided India during the colonial phase changed overnight into a peripheral state of the country, bounded by the then East Pakistan which later emerged as Bangladesh in 1971 as an independent nation.

During the colonial period the present state of Tripura was known as Hill Tipperah; mainly rural in character, ruled by the Manikya Dynasty of *Borok* kings (Bhattacharya, 1980). In 1949 Tripura was integrated with the Indian union as part 'C' state. In 1956, it became a centrally administered Union territory and later attained full-fledged statehood on the 21stof January 1972 under the North-Eastern Areas (Reorganization) Act of 30th of December 1971.

Present Tripura previously had dominance of tribal communities who lived in close proximity to nature and practiced shifting or slash and burn cultivation locally termed as *Huk* (Debbarma, 2009) or *jhum* cultivation. Even after integration with the Indian Republic, the tribes inhabiting the hilly terrain of the state , practice *jhum* (Gupta, 2000). Till recently *jhum* based livelihood practices provided ideal livelihood as the geographical and ecological conditions of Tripura made *jhum* cultivation an ideal mode of livelihood. Scholars (Dasgupta, 1986; Gupta, 1994; Gupta, 2000; Debbarma, 2009, Rahman 2006, Devi 2010) are of the opinion that the practice of *jhum* cultivation in the past was never a threat to the environment as the particular geo-environmental conditions favored such agricultural practice. The *jhum* cycle was longer helping in the rejuvenation of the forests cleared for *jhum* cultivation, the population size was also smaller and the small communities had access to plenty of land, forests and other natural resources for earning meaningful livelihood without causing much damage to the environment.

Dasgupta (1986) opines that shifting cultivators (*jhumias*) were not badly off even in late 1930s as *jhuming* provided them not only their livelihood sources but also a steady income from selling the surplus produced in the *jhum* fields. The *jhum* fields provided nutrition to the *jhumia* families as multi-cropping is common in *jhum*. The joy of *jhuming* for the *jhumias* slowly changed over time due to several factors. With Independence came modern intervention and development programs leading to decline in death rates but not adequate decline in birth rates paving in the natural growth of population which resulted to a steady increase in numbers, this increased the pressure of population on land and other natural resources resulting to reduction in the shortening of the *jhum* cycle, as the *jhum* cycles shortened , land degradation, deforestation and low productivity increased ensuing higher incidence of poverty mostly among the *jhumias*. Besides the natural increase of population Tripura also experienced large scale immigration. The partition of India and Pakistan led to the creation of East Pakistan (present Bangladesh) on religious lines and large scale Bengali speaking Hindu population migrated to adjoining Tripura (Dasgupta 1986, Dev Varman 1999, Gupta 2000) leading to population presure on the fragile resources of the state, soon land alienation issue emerged especially for the tribal communities (Dasgupta 1986) as about 50 percent of the *jhumia* population were severely affected (Dev Varman, 1999). Successive spikes of immigration from erstwhile East Pakistan were felt during the mid 1963 due to communal riot (Gupta, 2000) and another wave of influx in 1967 followed by the Bangladesh Liberation war in 1971 (Bareh, 2001). Consequently the tribal communities living here has been pushed towards the remote and hilly forested tracks of Dhalai and South Tripura districts where they still practice *jhum* cultivation (Debbarma, 2009), the plain areas are mainly occupied by the immigrants, hence the *jhumias* at present are mostly concentrated in the hilly and forested tracks of the state locally termed as *tillas* while most of the fertile plains (*lungas*) are occupied by the migrants practicing settled agriculture (Rahman, 2006).

Tripura like other states of India is on the path of development and various developmental projects like Gumati Hydro-electric Project have been introduced here leading to displacement of the local population, consequently most *jhumia* families report high incidence of land alienation and poverty as *jhum* cultivation is no longer sustainable leading to land degradation, deforestation and ecological imbalance (Dasgupta, 1986). Thus, practice of *jhum* cultivation has been discouraged by the administration.

It is interesting to note that from 1888 till the first half of the 20th century, the rulers of Tripura had attempted various *jhumia* rehabilitation schemes so as to encourage settled agriculture among the tribal communities (Debbarma, 2009); however these attempts failed (Administrative Report for 1939-40, Government of Tripura, Section 176-178, cited by Gupta, 2000; Debbarma,2009).

With Indian Independence and integration of Tripura with the Republic of India, the state government continued to formulate various schemes to control *jhum* cultivation, which dates back to first five year plan period of 1953, however, these schemes were not successful as it could not offer sustainable solution (Krishna kumar and Meenattoor.J, 2003; Maithani, 2005). 'The tribals could not eke out a proper living from such rehabilitation programmes' (Kumar and Meenattoor.J, 2003). Debbarma (2009) states that though many *jhumia* families have shifted to settled agriculture dominated by wet paddy cultivation; yet *jhum* cultivation is still being practiced widely throughout the state. Partition and shift of population from erstwhile East Pakistan, later Bangladesh has impacted not only the political scenario but also the socio-economic and cultural setup of the *jhumias* population. Bareh (2001) and Das et al. (2012) are of the view that one of the most important strategy of the state government adopted for *jhumias* rehabilitation along with the objective to provide a lucrative alternative to *jhum* cultivation in Tripura has been the introduction of rubber plantation in this state.

Rubber (*Hevea brasiliensis*) plantation in Tripura was introduced around 1963 by the Forest and Soil Conservation Department, Government of Tripura (Vinod et al , 1996). Rubber plantation was basically introduced in this state on a trial basis as part of afforestion programme for soil conservation measure in the degraded forests affected by *jhum* cultivation (Jacob, 2000 cited by Chaudhuri et al., 2013; Viswanathan and Bhowmik, 2014). The success of this trial motivated the Forest Department for more extensive cultivation of rubber trees (Chakravarti, 1986; Bhowmik, 2006; Raj et al., 2012) but rubber plantation being a new commercial venture had few takers and the tribal population out rightly rejected this new venture. However, some of the rubber trees that were planted during 1963 matured in the early 1970s which led to the commercial tapping and processing of rubber that started around 1971-72. Soon the local tribes realized the economic importance and ecological benefit of rubber plantation (Bhowmik, 2006; Viswanathan and Bhowmik, 2014). Consequently, the state government encouraged rubber plantation as it realized that rubber plantation can be useful in rehabilitating the *jhumias* and landless population into more economically lucrative and socially viable livelihood practices, resulting to the enhancement of their socio-economic conditions besides checking further deforestation and soil erosion caused by the practice of *jhum* (Viswanathan and Bhowmik, 2014).

Accordingly, the rehabilitation of the marginalized section of people through rubber plantation was encouraged which were initiated by different agencies (State government undertaking) *viz.* Tripura Forest Development and Plantation Corporation limited (TFDPCL) and Tripura Rehabilitation and Plantation Corporation Limited (TRPCL). At present the state government encourages the state tribal/caste Welfare Department in collaboration with the Rubber Board, the Tripura Tribal Area Autonomous District Council (TTAADC) and the Sub-divisional District Magistrate (SDM) and Block Development Officer (BDO) for initiating the rehabilitation of the marginalized section of people through rubber plantation (Bhowmik, 2006; Joseph et al., 2010; Viswanathan and Bhowmik, 2014). The rehabilitation through rubber plantation scheme has been implemented with the objectives to promote rubber plantation and bring the *jhumias* and other landless families mainly belonging to schedule tribal (ST) communities under settled agriculture by providing them with alternative livelihood options. This scheme was introduced with the objective to usher in sustainable economic development by improving the socio-economic condition of the landless marginalized population along with checking de-forestation and soil loss associated with jhum (Sarkar, 1986; Bhowmik, 2006; Raghavan, 2009; Debbarma, 2009; Choudhury, 2009; Viswanathan and Bhowmik, 2014).

Functioning of Rubber Plantation Scheme

Under the rubber plantation scheme each beneficiary household gets an allotment of one hectare (ha) land for developing rubber plantation. The beneficiaries earn income by selling the rubber latex or rubber sheets after a gestation period of seven years when rubber trees starts producing latex. The beneficiaries of the scheme are employed as wage labourers in the rubber plantation, for the initial seven years when land development and other production activities are taken up like inter-cropping of fruits like banana, pineapple etc for the market, this encourages alternative livelihood practices and income generation (Sinha, 2007, cited by Viswanathan and Bhowmik, 2014). The rehabilitation of the *jhumias* and other landless people through rubber plantation has been noted as one of the most promising schemes addressing rehabilitation of the *jhumias* and other marginalized landless people of thestate (Chakravarti, 1986; Gupta, 2000; Krishnakumar and Meenattoor.J, 2003; Maithani, 2005; Bhowmik, 2006; Raghavan, 2009; Devi, 2010; Sarkar, 2011; Raj et al., 2012; Das et al., 2012). Among the Northeastern states of India, Tripura at present has emerged as the most successful state in weaning away the ST from the practice of *jhum* cultivation to settled cultivation, providing livelihood security through rubber plantation for the past four decades (Sharma et al., 2011). Rubber plantations has emerged as an important source of economic dynamism helping more than 57,000 households and the economy of the state (Directorate of Economic and Statistics Planning Department (DESPD), 2015; Bhowmik and Viswanathan, 2015).

Tripura has emerged as one of the non-traditional rubber growing states of the North-East India and the area under rubber plantation here has been expanding tremendously. This can be attributed to favorable agro-climatic conditions coupled with low elevation and soil composition (Gupta, 2000; Maithani, 2005), terrain suitability and acceptability amongst the local people (Devi, 2010; Sarkar, 2011). As per the Rubber Board during 2015-16 the area under rubber plantation in this state is 74,335 ha accounting for 7.09 percent of its total geographical area. Tripura stands second both in area under rubber plantation and production of natural rubber (NR) in India next only to Kerala. Consequently, Tripura has been declared as the 'Second Rubber Capital of India' by the Rubber Board (Devi, 2010; Viswanathan and Bhowmik, 2014; Rubber Board,2016).

The momentum for commercial value of rubber plantation in Tripura is very recent that started around mid-1980s with rubber being identified and accepted as one of the most commercially viable cash crop by the people of Tripura. The basic objective of this scheme are mainly to rehabilitate and bring the *jhumias* and other landless ST people within the ambit of settled agriculture and offer them with diverse livelihood opportunities, so that they can improve their socio-economic conditions. However, the question arises as to whether the existing rubber plantation programme has really achieved its goal in rehabilitating the *jhumias* and other landless ST families by offering them with alternative livelihood options thereby enhancing their socio-economic development. Previous studies on the impact of rubber plantation based rehabilitation of the marginalized section of population *i.e.* the tribal communities suggest two distinctively opposite views *viz.* one group of scholars are of the view that rubber plantation has been successful in providing multiple livelihood options to the marginalized ST population thereby enhancing their socio-economic condition while the other group of scholars are of the opinion that rubber plantation has not been much successful in providing rehabilitation and associated socio-economic opportunities to the marginalized ST population. Thus, the question arises as to whether the benefits of rubber plantations are really reaching the communities for whom this scheme has been introduced. It is in this context the present book evaluates the growth and distribution of area under rubber plantation in the state of Tripura and the associated livelihood options paving in socio-economic development of the marginalized population.

History of Rubber Plantation in the World

Hevea brasiliensis, which is commonly known as rubber tree belongs to the *Eupohorbiaceae family*. Rubber trees (*Hevea brasiliensis*) are considered important due to its commercial value as it produces high quality latex (main economic product which is processed into rubber sheet) than any other genus hevea plants species which grows widely in the native Amazon region (Mors and Rizzini, 1966; Opeke, 1982; Peter, 2002; Gleason, 2005). Polhamus (1962), Mors and Rizzini (1966) states that the success of rubber plantation is credited to an Englishman –Sir Henry Alexander Wickman (1846-1928) who smuggled *Hevea brasiliensis* seeds out of Amazon rain forest in 1876 and introduced it in the Royal Botanical garden at Kew. This marks the starting point in the distribution and dispersion of rubber trees all around the tropical regions of the world. The British introduced rubber plantation to feed the highly lucrative imperial rubber industry. The first rubber plantations were introduced by the British in the island of Ceylon i.e. present Sri Lanka and later it was introduced to other European colonies in the far-east (East Asian countries) for commercial gains here large tracks of tropical forests were cleared by the native labour who provided cheap labour. The demand for rubber increased with the growth of the American – motor car industry where the car tyres were made of rubber. Over time the demand for rubber grew exponentially and the market for plantation rubber was assured.

Polhamus (1962) states that since the establishment of rubber plantation through out the tropical region in the early part of 19th century, rubber plantation became the chief means of livelihood in many countries of Far East, like Indonesia, Malaya (Malaysia), Burma (Myanmar), Ceylon (Srilanka), Vietnam, Cambodia, Thailand, Brunei, Sarawak and British North Borneo (Borneo). Millions of people of these countries dependent directly and indirectly on the wages from rubber plantation and production of natural rubber. Vast tracks of tropical forests were cleared for setting up rubber plantation to feed the rubber industries of Great Britain. The Imperial rubber industry was set up on exploiting native population which provided cheap labour and the colonies lost their biodiversity along with the forests due to large scale rubber plantation.

This scenario changed after the Second world war, as the map of the world changed and a new political order was established, the previous colonies emerged as independent nation states asserting their rights ushering in development programmes so as to bring in a more equitable distribution of resources and inclusive development of the society, with it , the role of rubber plantation also changed. The rationale behind rubber plantation was no longer exploitation of cheap labour but to provide livelihood to the marginalized in the society. As per the Food and Agriculture Organization (FAO) of the United Nation (2013) about 10 million people are earning their livelihood from collecting latex from rubber trees either as a full time job or as a major livelihood activity second only to farming as their main source of livelihood. Most of these people are from South and Southeast Asian countries i.e. Thailand, Indonesia, India, Malaysia and Sri Lanka. Previous studies indicate various issues are emerging in realizing inclusive development through rubber plantations which are discussed at global, national, regional and local levels.

Global Level

Ruthenberg (1980) attempts a comparative study between large scale rubber plantations mainly located in Malaysia, Sumatra and Sri Lanka with small producers from Sri Lanka, Sumatra, Malaysia and Nigeria. Rubber being an important cash crop is extensively grown by both large rubber estates and smallholders; however, the objectives and techniques of production differ between the large estates and small producers. Production of NR in estates are mainly for maximizing of financial returns while among the smallholders it is considered as part of an integrated farming system where rubber serves as a supplementary source of income. Ruthenberg highlights on the point that as tapping of rubber trees is not a regular task, smallholders prefer to tap rubber trees when time permits or when cash is needed. Another advantage for the-smallholders' in rubber plantation is – the employment of farm labour at low costs. It is observed that when rubber estates are severely hit by falling price in the market, smallholders tend to cultivate other crops more intensively and the rubber trees are kept for rejuvenation to produce larger yields when rubber prices rises.

Opeke (1982) states that the favorable agro- climatic conditions of the tropical and sub-tropical world attracts rubber plantation, due to this rubber plantation have emerged as a mono crop in the tropical and sub-tropical parts of the world including India which has brought about land cover land use changes .Liu, Hu, Ma and Li (2006) demonstrates the land use changes brought about by expansion of rubber plantation by using Landsat Thematic Mapper (TM) image of 1998 and Landsat Enhanced Thematic Mapper (ETM) image of 2003 with 30 meters (m) spatial resolution. They further link the land use land cover (LULC) changes in rubber plantation areas with changes in the socio- economic conditions of the villagers. Their findings suggests that expansion of area under rubber plantation has been rapid which has increased from 4,042 ha in 1998 to 13,104 ha in 2003 suggesting an increase to the tune of 324 percent. The scholars are of the opinion that increase in area under rubber plantation has been at the expense of forested area, fallow fields and shifting agriculture. Most of the expansion has taken place in the lowland areas having suitable climatic condition which are close to roads. From 1998 to 2003, the villagers/rubber grower's income has increased due to rubber plantation. These suggest that the economic benefits from rubber growing are likely to continue resulting to the expansion of rubber plantation. The massive expansion of rubber plantation has been promoted by a series of policy intervention such as the Green Industry Policy along with the Sloping Land Conversion Program, in order to eliminate swidden/shifting agriculture and promote cash crop rubber plantation so as to increase forest coverage. The scholars point-out that though adoption of rubber plantation has enhanced the economic benefits of the villagers, yet it needs to be viewed carefully over a longer period of time as a collapse of rubber market could have disastrous consequences over these communities, as the market is privatized and the government is lacking in protecting such growers which they were enjoying in the past. Consequently, the authors of this paper are of the view that in order to promote sustainable development, the local people and the government should cooperate and monitor carefully the possible effects of the so called 'Green Industry', minimizing the expansion of rubber plantation and protecting local forest.

Viswanathan (2008) has attempted a comparative analysis on the performance of rubber farm-based livelihood systems of small holders, this is based on case studies of two regions viz. Northeast India and Thailand. The conceptual framework of sustainable livelihood analysis (SLA) developed by the Department for International Development (DFID) has been used for analyzing the impact of rubber farming system. He is of the view that in both the countries rubber plantations are viable provided prices remain remunerative in the future. His finding indicates that rubber production not only contributes to the household income but also to the economy of the states. The sustainable livelihood perspective and socio-economic significance of the integrated rubber farming systems as introduced in Northeast India with the rationale to rehabilitate and improve the socio-economic conditions of the marginalized section of the ST population needs further probing as the small rubber growers are highly vulnerable to market uncertainties. A livelihood is sustainable when it can cope with and recover from stress and shocks, maintain or enhance capabilities and assets without undermining the natural resource base. It is in this context that the integrated rubber plantation system provides the smallholders with ample capability for resilience during crises and ensures sustainable flow of income for such households. He addresses the five assets/capitals associated with sustainable livelihood viz. (i) Natural capital, (ii) Human capital, (iii) Physical capital, (iv) Economic or Financial capital, and (v) Social capital and measures it by taking suitable indicators representing the five assets/capitals. The enhancement of especially financial capital brings changes in socio-economic setup of the rubber growers in Northeast India and Thailand. He concludes that in Northeast India there is need to strengthen the smallholders' access to different forms of assets/capitals as in Northeast India the access to secure property rights over area under rubber trees are restricted. This calls for policy and institutional intervention in order to secure the appropriate property rights for the small holders.

Pensuk and Shrestha (2008) attempts to examine the effect of land use change, in particular paddy to rubber impacting rural livelihoods in Phatthalung Watershed, Southern Thailand. The authors have used Landsat images of three time periods

(1976, 1990, and 2006) and ERDAS IMAGINE software to interpret the land use changes. In order to analyse livelihood status, sustainable livelihood approach has been used taking the five different assets/capitals *viz.* human, natural, financial, social, and physical assets/capitals. Their findings indicate that there has been a dramatic decline in rice growing area due to expansion of area under rubber plantation as many rice growing farmers have converted their land into rubber plantation due to the higher return from rubber production. The authors are of the opinion that rubber growing households tends to be better-off in terms of financial asset/capital due to generation of more income, whereas households practicing paddy are better-off in physical asset/capital due to their easier access to market. Human, natural and social assets/capitals are not significantly different between the two groups of households viz. paddy growers and rubber growers.

Fu, Brookfield, Guo, Chen, Chen and Cui (2009) are of the opinion that, due to increase in income from rubber plantation, rubber plantations are regarded as a principal indicator of wealth status among the households, consequently leading to livelihood improvement of the smallholders. The increase in price of natural rubber after 2002 have decreased the number of cultivated upland rice varieties from seven in 2001 toonly one variety by 2004. Consequently, area under rubber plantation has been increasing from 17.7 ha in 1998 to 82.2 ha in 2004 on the other hand swidden/ *jhum* fields have decreased from 20.4 ha in 1998 to 12.7 ha in 2004. The authors highlight that besides mono culture farming of rubber; several rubber related agro-forestry systems are also being practiced by local smallholders, as this generates diversified livelihood practices generating higher income for the individual/ households . This leads to inequality of income among households, primarily due to variations in income from rubber. Rubber growers have shown livelihood improvement especially among the small holder households leading to transformation from traditional subsistence agriculture to commercial agriculture based on growing high-value cash crops. Due to high returns from rubber plantation most of the households are progressively altering their traditional land-use pattern and labor allocation towards a growing cash crops, overwhelmingly rubber plantation.

Verheye (2010) in his article 'Growth and Production of Rubber' highlights that rubber trees have been cultivated even at 25 degree North of Equator in Yunnan Highlands of China and 21 degree South of Equator in Brazil. He however states that most of the rubber plantations are concentrated between 15 degree North and South of Equator in Southeast Asia and South Pacific countries which includes Malaysia, Indonesia, Thailand, Sri Lanka, India, Cambodia, Vietnam, Philippines, Papua-New Guinea and Southern China, Central and West Africa countries of Congo, Cameroon, Ivory Coast and Liberia, rubber plantations has also been setup in tropical America however concentrated over a small area within 10 degree North and South of Equator.

Fox and Castella (2013) in their research article opines that in mainland Southeast Asia, the traditional rubber growing areas include portions of southern Thailand, south-eastern Vietnam, and southern Myanmar. However, the success of rubber plantation of the Chinese government in the non-traditional environment that lies north of 22 degree North latitude. Many of the entrepreneurs from China, Vietnam, Malaysia, and Thailand are investing heavily on the expansion of rubber plantation in non-traditional areas of Laos, Cambodia, Myanmar, in northwest Vietnam and northeast Thailand. In Thailand rubber plantation has expanded and includes over 64, 000 ha of land in the north and 348, 000 ha of land in the northeastern parts. In Vietnam rubber plantation has increased from 395, 000 ha in 1999 to 550, 000 ha in 2007 with the government targeting 700, 000 ha of land to be brought under rubber plantation by 2020. The authors are of the opinion that in Xishuangbanna, Yunnan, China and northeast Thailand rubber production by smallholders are viable and is an effective proposition for moving households and communities out of poverty. However they also highlight that many of the farmers in Laos, Combodia and Myanmar are struggling to maintain their land and forests in the face of growing pressures from investors and government institutions to impose concession arrangements.

Luangmany and Kaneko (2013) reviews the factors that has accelerated rapid expansion of rubber plantation in northern Laos and its impact on economy and environment.. According to their review the Land and Forest Allocation program has played an important role in shaping the landscape of northern Laos. The authors have also observed that the increase in the demand of natural rubber from the Chinese market and the increase in the global price of natural rubber are the most important factors for the rapid expansion of rubber plantation in northern Laos. They are of the opinion that in northern Laos investment in small holder production of natural rubber is considered economically profitable under the current situation which is likely to continue in the future in spite of fall in food production, soil quality degradation and deforestation. They suggest that the behavior of small holder rubber production in northern Laos need thorough analysis so to understand their 'labor allocation decisions for different livelihood activities, including forest extractions in order to achieve the goal of promoting rubber tree plantation as a key strategy to alleviate poverty as well as an instrument to eliminate shifting cultivation in order to conserve natural forests'.

Nath, Inoue and Zoysa (2013) examine the enhancement of people's livelihood through small-scale rubber plantation in Bangladesh, India, and Sri Lanka. They have adopted and used the five livelihood assets/capitals (human, natural, physical, financial, and social assets/capitals) of the sustainable livelihood framework to analyze the impact of rubber plantations on the rubber grower households'. Their finding indicates that in all the three study sites i.e. Khagrachari hill (Bangladesh), South Tripura (India) and Kamburupitiya (Sri Lanka) rubber plantations has brought positive impact on the livelihoods of the rubber grower households, which differs among the three site. Households income from rubber plantation combined with other sources are found to be sufficient for house hold food requirement in South Tripura and Kamburupitiya in comparison to Khagrachari hill. The institutional support such as technical and financial support from the Rubber Board in India and Sri Lanka has encouraged the rubber growers to managed the plantation properly resulting to better stock, growth and management than those of Khagrachari hill in Bangladesh. In Khagrachari hill there is absence of rubber advisor board/rubber research organization to support the rubber growers. Due to the income derived from rubber plantations, households receive loans from non-government organizations (NGOs) for other activities like agriculture, small businesses, and livestock rearing, this helps in diversifying their livelihood strategies to cope with the any crises. The rubber growers stated that although rubber plantation can lead to diversification of livelihood practices, at present their income is at risk as it relies on single commodity (rubber) whose price fluctuates as per market conditions.

Ahrends, Hollings worth, Ziegler, Fox, Chen, Su and Xu (2015) analyzed the expansion of rubber plantation in continental Southeast Asia between 2005 and 2010 in combination with the environmental data and the performance of rubber plantation. Their findings suggest that rubber plantations are spreading into sub-optimal/marginal environments thereby resulting to unsustainable growth. Rubber plantations have been spreading at the expense of forested areas and between 2005 and 2010 more than 2,500 km2 of natural tree and 610 km2 of protected areas have been destroyed for growing rubber plantation. The authors' highlights that expansion of rubber plantation in the marginal areas creates potential loss of high-biodiversity, altering the land use land cover thereby resulting to unsustainable growth. The existing livelihoods gets compromised due to fall in the price of natural rubber. The authors are of the opinion that appropriate policy interventions and awareness are an urgent need for systematic and region-wide monitoring of rubber plantation so as to quantify the losses and impacts of expansion of rubber plantation into forest area and ecosystem so as to limit further environment losses as well as societal impacts.

Chen, Yi, Schmidt-Vogt, Ahrends, Becks chafer, Kleinn, Ranjitkar and Xu (2016) analyzed the spatial pattern and dynamics of the expansion of rubber plantation in Xishuangbanna, South West China for the years 1988, 2002 and 2010. They used Landsat imageries of 30 m resolution for the years 1988 and 2002 and Rapid Eye with 5 m resolution for 2010 and mapped using Arc GIS 9.3. Their findings suggest that area under rubber plantation has been expanding. 10 percent of the total forest area within Xishuangbanna has been converted to mono culture rubber plantation by 2010. They conclude that expansion of rubber plantation has been extremely rapid from 1988 to 2010. Rubber plantations are expanding towards the higher elevations, steep terrain and into forest tracks; this has created not only serious threat to biodiversity and environmental services but also decline in the production of natural rubber. The authors suggests that local government need to take essential steps to develop long-term land use strategies so as to balance economic benefits from rubber plantation along with environment sustainability is also the need to assists the farmers with proper knowledge for the selection of land suitable for rubber plantation.

Vongkhamheng, Zhou, Beckline and Phimmachanh (2016) examine the impact of rubber cultivation on the socio-economic and ecological set up of Southeast Asia. According to the review they noticed that rubber plantation has high commercial value and is an income generating activity which is good for rural communities as it ushers in socio-economic development. The expansion of rubber plantation in order to meet the global burgeoning demand for natural rubber is however destabilizing rural livelihoods, reinforcing the poverty nexus, harming watersheds and destroying tropical forest ecosystems. However, with adequate and effective management and proper sustainable planning along with involvement of stakeholders, governments and rural communities in the decision making process, the cultivation and establishment of rubber plantation can have a very positive socio ecological impact.

National and regional level

George, Haridasan and Sreekumar (1988) examines the structural changes in rubber plantation industry since the 1950s and analyzes the major role of government in the development of this industry. They highlight that it was between 1960-61 and 1965-66, that the major increase in area under rubber plantation took place in India mainly catering to the estate sector due to the emergence of public sector corporations. The public sector corporations have taken an active interest in planting rubber on a large scale under the initiative of the government of India and the concerned state governments. In Kerala, Plantation Corporation (established in 1962) took over the rubber plantation, which was set up initially by the state government resulting to the expansion of area under rubber trees. Some private companies converted their large tea estates into rubber plantations which are mainly reported in southern Kerala. In Tamil Nadu and Karnataka, Forest Departments of the respective States have taken initiatives in planting rubber trees on a large scale. In Assam and Tripura, rubber plantations are undertaken mainly by the public sector corporations. In Meghalaya and Mizoram, the state soil conservation Departments and in Manipur the forest department is involved in rubber plantations. Individual entrepreneurs are also showing interest in rubber plantation in the Northeastern states of India. The scholars are of the opinion that due to the initiatives of the governmental agencies in planting rubber trees, there have been changes in the geographical distribution of area under rubber as well as changes in the ownership pattern of rubber plantation both in the traditional and non-traditional areas of rubber plantation.

Burman (1990) in his paper 'Shifting Cultivation in the Hills of Assam' states that rubber plantation has not been very successful as it could not bring about a change in the socio-economic conditions of the *jhumias* of North-East states of India i.e. from *jhumias* to owners of rubber plantation as some of the *jhumias* are only engaged as wage laborers in the rubber plantation that are often owned by the elites of such societies. The scholar states that the expansion of rubber plantation in the hills of Northeast India in the long run may help in improving the socio-economic conditions of the marginalized section comprising of *jhumias* (who are at present working as wage laborers in such rubber plantation) if policy intervention and planned measures are introduced by the government through proactive planning.

Bordaloi (1990) highlights that the major aim of the Assam Cash Crop Plantation Development Corporation (setup in 1974) are to promote rubber plantation and usher in economic development as well as to rehabilitate the *jhumias*. He believes that if the scheme and programme are implemented with patience and enthusiasm along with the support of the community leaders' success is inevitable.

Borthakur (1992 highlights that the main objective in introducing rubber plantation in the Northeastern region is to provide alternative livelihood opportunities to the *jhumias* by slowly incorporating them within settled agriculture, generating employment in such plantation as well as encouraging conservation of the degraded forest areas and checking soil erosion associated with *jhum* cultivation.

Kurien (2001) analyzes the livelihood aspects of marginal farmers covered by the project- 'Rubber to the Poor' in the state of Kerala. In Kerala, Rubber to the Poor Project was implemented in the early part of 1980s by the Malanad Development Society (MDS) in collaboration with the Rubber Board. The main goal of the project was to extend help to the smaller landholders to grow rubber trees so that they can earn an additional income and improve their socio-economic status. The Rubber Board provides technical guidance/advice and instruction for the scientific cultivation of rubber trees to the beneficiaries. The MDS appoints local supervisors who gives technical skill to the rubber growers The MDS also provides planting material, fertilizers, copper sulphate and calcium carbonate necessary for such plantation. Training regarding production and processing of rubber are provided by the MDS unit secretaries and supervisors. The finding indicates that the beneficiaries have shifted their major livelihood practices from marginal farmers to latex tappers. Women are involved in large numbers in nursing rubber trees and harvesting the yield. After the rubber trees mature, tapping is usually done by professional private tappers as well as by the owner of the beneficiary household. The beneficiaries' who are also latex tappers can complete tapping in their respective plantation farm and can also take up casual work elsewhere the same day thereby increasing their income. After tapping, latex is collected, the rolling of the latex into lump sheet (processing) is done followed by thinning of sheets etc which are mainly done by female members of the beneficiary house hold.

Viswanathan and Shivakoti (2008) examines the influence of important socio- economic, institutional/policy level factors in determining the adoption of rubber- integrated farm-livelihood systems in traditional rubber growing regions of Kerala (South India) and the non-traditional rubber growing regions of Assam, Meghalaya and Tripura (Northeast India). There is a contrast in the adoption of rubber-integrated farm-livelihood systems between the two rubber growing regions *i.e.* Kerala and Northeast India. Socio-economic factors such as shrinking of land holding size, non- availability of family labor and the rubber mono culture oriented policy promotional schemes offered by the institutional intervention of Rubber Board has acted to a great extent in the limited adoption of integrated rubber farming systems in the traditional regions of Kerala. In Northeast comprising of the emerging rubber economics these integrated rubber farming systems are welcome by the tribal communities. The institutional support and Research and Development (R&D) intervention of the Rubber Board in implementing cultivation of rubber trees on a commercial scale in the late 1980s has played an important role in expansion of rubber plantation in North-East India. The rubber plantations in

North-East India are established in barren and marginal lands, an overwhelming majority of the prospective rubber growers follow an integrated and diversified farm-livelihood system comprising of upland paddy, fishery, poultry, livestock, annual/perennial crops, piggery etc. One of the most explicit positive impacts of the integrated farm- livelihood system and rubber growing in North-East India is the continuous flow of income from rubber plantation that has enabled the tribal communities to avoid the situations of distress sale of paddy which they did previously, before adopting the rubber plantation scheme. Unlike the traditional regions, North-East India also offers scope for expansion of rubber-integrated farming system as a measure of sustainable agriculture in the era of globalization. The authors suggest that new institutional paradigm must be corroborated with specific R&D intervention to evolve sustainable rubber-integrated farm livelihood system for the North-East India without compromising the socio-economic, ethnic and ecological diversity of the region.

Rajasenan (2010) in his report 'Livelihood and Employment of Workers in Rubber and Spice Plantations' compares the livelihoods based on rubber and spices plantations in the state of Kerala. For assessing the livelihood assets of the plantation worker he adopts the conceptual framework developed by USAID and ILO and classifies the livelihood assets into physical, natural, financial, human, social, political, and welfare assets. In the state of Kerala plantation crops has been providing livelihood opportunities to about 14 lakhs families and has helped in developing many of the remote areas into rubber, coffee, tea plantation etc. Among the plantation crops, rubber plantation provides the maximum number of employment opportunities thereby providing livelihood to a large section of population. Based on the findings of the seven livelihood assets of the rubber and spice plantation workers Raiasenan concludes that the livelihoods of the workers involved in rubber plantation are better off than the workers of the spices plantation as their livelihood options are more vulnerable in comparison to livelihood opportunities provided by rubber plantation.

Nath and Bezbaruah (2010) attempts to examine the profitability and economic sustainability of rubber production in Goal para district of Assam where the annual compound growth rate of acreage under rubber plantation has grown at a high rate to the tune of 17.5 percent during 1986-87 to 2006-07 while the annual compound growth rates of production and yield were 19 percent and 2.9 percent respectively during 1986-87 to 2006-07. Their finding indicates that rubber growers have diversified and enhanced their livelihood options besides changing their respective per-existing occupations. They conclude that the growth and production of rubber in their study area has emerged as an important economic activity and majority of the growers are from socio-economically disadvantaged background. Rubber plantation has served the cause of economic empowerment at the grassroots. The authors are of the opinion that rubber plantation has provided the rubber growers with alternative livelihood options thereby helping the growers to cope with the risk associated with other commercial activities. The gain in income from rubber production has led to better purchasing power leading to better access to food for the rubber growers.

Kannan (2013) in his study 'Growth and performance of area, production and productivity of natural rubber in India' analyze the growth and performance of natural rubber in India and major rubber producing states in India from the period 1981-82 to 2010-11 using compound growth rate and co-efficient of variation. The analysis reveals that the growth in both production and yield of natural rubber in India registers a significant positive growth. The co-efficient of variation indicates that there is not much variation in case of area under mature rubber for tapping, area under rubber plantation and yield as compared to the production of natural rubber in India. However the growth in area of natural rubber in traditional rubber producing states of India (Kerala, Tamil Nadu and Karnataka) suggests a declining trend though productivity has increased.

Vongkhamor et al. (2007), Food and Agriculture Organization (FAO) of the United Nation (2013) and Lagerqvist (2013) states that over a million of Indian households are directly dependent on rubber plantation (on NR processing and product making sectors) for their livelihood. Most of these families are smallholders of land and they generally tap rubber as a major livelihood option as they tap the latex from the rubber trees at the same time till their paddy fields and usually collect the latex on their way home.

Meti, Pradeep, Jacob, Meerabai and Jessy (2014) has attempted to examine the spatio-temporal changes in area under rubber plantation in Kanya kumari district of Tamil Nadu using remote sensing and GIS techniques over a period of 27 years (1980-2007) and its distribution over soil management units (SMU) and slope. They have used Landsat Multi spectral Scanner System (MSS), TM, and Indian Remote-Sensing Satellite (IRS) P6 Linear Imaging Self-Scanning Sensor (LISS) III for acquiring satellite images and ASTER digital elevation data for deriving slope information and or-tho-rectification of satellite image. Their finding reveals that rubber area has substantially increased from 10,693.3 ha in 1980 to 20,781.7 ha in 2007 in Kanyakumari district. The overlay image of area under rubber on the slope and SMU map indicates that majority of rubber area are distributed over good to moderate soil with slope varying between moderate to gentle. They suggest that there is a need for integrated analysis of rubber area distribution as per soil condition, topography, and climate variation so as to assess the influence of these factors on rubber performance as well as develop site specific management practices.

Geosrge and Chandrashekar (2014) attempted to analyze the growth of area, production and productivity of rubber and also examine the marketing of natural rubber(NR) by different producers in Kerala. For analyzes, the compound annual growth rate (CAGR) technique has been adopted. Kerala ranks first in the production of NR in India. In 2011-12, area under rubber plantation in Kerala was 5, 39,565 ha and production was 7, 88,940 tonnes while productivity was as high as 1,462 kg/ha. Major producers of NR in Kerala are Kottayam, Ernakulam and Pathanamthitta districts. From 2000-01 to 2011-12, the CAGR of area under rubber plantation in Kerala was 1.309 percent and CAGR in production of NR was 3.097 percent while productivity was 1.750 percent. Marketing system of NR by the small holders is different from the estate sector. Small holders' sell rubber sheets and latex through village level dealers or merchants while in the rubber estates various facilities for processing different types and grade of rubber are sold through proper tendering and auction.

Local Level

Gupta (1990) highlights the economic benefits of the rubber plantation-based rehabilitation programme on the *jhumia* beneficiary households. The economic benefits appears significant for short term but does not seem to be too hopeful in the long term. Most of the *jhumia* beneficiary households lack in work ethics and have poor technical knowledge which is necessary for maintaining the plantation. They are more satisfied to work as wage laborers than becoming owners of such plantations. Further the land under rubber plantation being mortgaged to the agency (government undertaking) acts as a deterrent to motivate the *jhumias* to look after the plantation land as their own. He notes that transition from *jhumias* to plantation workers is possible, but is not that easy in practice as it often becomes difficult for the *jhumias* to quickly transform themselves from *jhum* cultivation to owners of plantation. Bhattacharya (1992) is of the opinion that *jhumias* are slowly accepting the modern way of earning income through tapping the rubber trees. During the immature phase of rubber trees members of the beneficiary households work as daily wage earners in the plantation instead of working in other places for earning their livelihood. After maturity of the rubber trees, different activities like development of homestead area and utilizing the same plantation area for other agri-crop, horticulture, social forestry, poultry, piggery etc has been initiated as livelihood options. Besides generating higher income, rubber plantation has opened up a new horizon of economic activities as the beneficiaries have become accustomed to latex collection (tapper). The rubber plantation also provides fuel wood, vegetable-oil, oil-cakes and honey as bi-products. In spite of sincere efforts by the government in rehabilitating the marginalized *jhumias* through rubber plantation, a good number of landless *jhumias* are not spontaneously accepting this project and are adverse to such scheme, thus the transition of *jhumias* from *jhum* cultivation to tapping of rubber trees is beset with various obstacle which needs careful consideration. Bhattacharya suggests that emphasis should not only focus upon mono culture plantation but also should highlight on the three-dimensional forestry based livelihood options. The administrative lacunae needs to be removed and education facilities needs to be expanded, with widening of the occupational pattern the *jhumias* will have a higher chance to accept the new scheme for improving their socio-economic conditions.

Maithani (2005) opines that rubber plantations are being transferred to the non- tribals and the elite section of the ST population residing in the state of Tripura through Benami/unauthorized transactions. He also states that rubber plantation scheme will ultimately skip the *jhumias* due to various inbuilt legal and procedural hassles, lack of food component in the scheme, highly regimented laborers and the lurking fear in the minds of the tribal's from their past experience that someday the government may take over the land allotted to them which is earmarked for rubber plantation. The government needs to take appropriate steps to encourage the marginalized ST population to realize that rubber plantation can enhance their socio- economic conditions.

Bhowmik (2006) suggests that there are various factors responsible for expansion of area under rubber plantation in Tripura, the introduction of cash subsidy by the Rubber Board, economic profitablity, rehabilitation programme of the *jhumias* by different state run agencies, increase in price of NR are some of the major factors for the development of rubber plantation in the state. Besides public sectors, rubber plantation has been promoted by private sector mainly due to profit maximization. Rubber plantation has diffused throughout the districts with highest area being concentrated in West Tripura district, followed by South Tripura, North Tripura and Dhalai districts. Bhowmik has also conducted field survey in two rubber plantation centers managed by TRPCL i.e. Bangshibari and Herma in order to assess the economic viability of rubber plantation for the rehabilitated beneficiaries. The analysis reveals that rubber plantations are economically viable and provides substantial income to the beneficiaries. However differences in income levels exist among the beneficiaries due to variation of landholding size of rubber plantation and production of latex, which is aptly reflected by the living standards and possession of valuables and assets by the beneficiary households. Bhowmik also highlights on the various problems/constraints to rubber plantation, as well as the prospects for rubber plantation with some suggestions for the development of rubber plantation in the state.

Sarkar (2011) analyzes the growth of rubber plantation and the income derived from rubber plantation. His findings indicate that area under rubber plantation is increasing rapidly. Different state government agencies (TFDPCL and TRPCL) and the Rubber Board have contributed positively in the development of rubber plantation helping in rehabilitating the tribal *jhumias* through rubber plantation. The tribal people can earn income from both immature plantation (through inter-cropping and casual worker) and mature plantation (through tapping, rubber agency, rubber industry, rubber wood, rubber seed, rubber honey, firewood, biogas and casual worker). Once the rubber plantation matures the tribal population can earn a higher income paving in development. Rubber plantation provides ample scope for employment opportunities as well as reduces deforestation and therefore the state government should initiate more rubber plantation for reducing unemployment in rural areas in particular and the state in general.

Raj *et al.* (2012) has identified and mapped the potential areas in Tripura for the expansion of rubber cultivation. Using satellite images to map area under rubber plantation, 48,037 ha of land is estimated to be under rubber plantation in March 2012; however this did not include the young rubber plantation of 2/3 years old as these young plantations often has similar spectral signature with other fallow lands/scrub lands making it difficult to map. Therefore, they added the total area under young rubber plantation of two years old which is 10,600 ha as per the Rubber Board data, and the total area under rubber plantation was 58,637 ha which closely corresponds to the official data of the Rubber Board i.e. 59,285 ha during 2012. Spatial distribution of area under rubber plantations are mostly concentrated in the western part of Tripura .Analysis through satellite data reveals that about 28,133 ha of potential wasteland areas like scrub land, degraded land, marginal and *jhum* land has been identified for further expansion of rubber plantation in the state. However about 5,186 ha of potential area for rubber plantations has been observed within the reserved forest area and 22, 947 ha has been noticed outside the reserved forest area where rubber plantation may have been already existing or can be planted. They suggested that rubber plantation should be actively promoted in the wastelands (outside reserved forest area) as rubber trees can protect further degradation of denuded wasteland and restore such ecosystem; thus rubber plantation expansion here can occur without interfering with the local food security and biodiversity.

Mohana kumar (2013) attempted to assess the role of expansion of rubber cultivation in Tripura and its impact on landholding, employment and income. He highlights that in Tripura the expansion of area under rubber plantation has gained its momentum from 1980s and the Gross Cropped Area (GCA) has increased from 3.05 percent in 1989-90 to 10.08 percent in 2009-10. Rubber plantation as a commercial crop was promoted vigorously by the Rubber Board and the government of Tripura from 1960s to 1980s and by 1990s rubber plantation covers substantial area and is the most remunerative crop in the state. Among all the districts, rubber plantations are mostly concentrated in the South Tripura and West Tripura districts of the state and accounts for more than 90 percent in 2012; however, rubber plantations are more developed in South Tripura district. He further highlights that as per the Rubber Board, the area under rubber plantation in the state is expected to expand by 1 lakh ha during the 12th and 13th five year plan period. Due to the increasing demand of natural rubber in the National and International markets the areas under rubber plantation are expanding faster than any other crops in Tripura. Tripura an agrarian state has been undergoing a process of transformation from food producing subsistence economy to rubber plantation based commercial crop production which is a matter of serious concern with respect to the economically and socially vulnerable sections of the society. He suggests that such deleterious consequences can be partly resolved by designing an agriculture policy including land use which can maintain the balance among income, employment, and food security of different categories and classes of people in the state.

Debbarma (2013) attempted to analyze the impact of rubber plantation on the socio-economic conditions of shifting cultivators in Dhalai district of Tripura. Taking the five assets/ capitals associated with sustainable livelihood *viz.* (i) Natural capital, (ii) Human capital, (iii) Physical capital, (iv) Economic or financial capital, and (v) Social capital and measures it by taking suitable indicators representing the five assets/capitals, his finding indicates that in Dhalai district the socio-economic conditions of the beneficiaries have substantially improved. The rubber growing households have gained substantially in their physical and financial assets/capitals. Their natural capital has slightly declined as their access to community land and forests are decreasing, however their human capital have slightly improved but they are slowly giving up traditional skills as the beneficiaries are in the process of learning modern skills. Regarding social capital the tribal communities living in Dhalai district have an inbuilt social capital as they are closely tied on the basis of kinship and are thereby rich in social capital. He concluded that at present a large section of the youth are employed in the rubber plantation as wage laborers or permanently employed as processors of latex in the rubber processing centers. This has paved in a social change in the area and opened up opportunities for the erstwhile *jhumias* and other landless families to improve their socio-economic conditions. He also noted that among the districts of Tripura, Dhalai district which has the highest concentration of ST population of about 54 percent to its total population has the least area under rubber plantation (6.43 percent) compared to West Tripura district which has about 46.72 percent of the total area under rubber plantation and has a share of only 25.3 percent of ST population to its total population, for whose upliftment rubber plantation has been introduced in the state. Mohana kumar (2014) highlighted that expansion of area under rubber has been rapid due to increase in demand and price of natural rubber in the domestic and international markets. The state of Tripura is experiencing a transition from rice producing subsistence state to commercial base, natural rubber producing state due to higher economic return from natural rubber. Area under rubber plantation has been increasing as both central and state government's policy intervention promoting rubber plantation. Most of the rubber plantations in the state have come up in West Tripura district (in south and central parts) and South Tripura district. Rubber plantation in North Tripura and Dhalai districts of Tripura has been marginal as most parts of these districts are covered by hilly terrain with forested areas. Rubber being a commercial crop, changes in price of rubber can result in virulent ups and down in the purchasing power of the people when compared to rice. Rubber plantation has also tilted the distribution of land, favoring rural rich including both tribal groups, while there has been a sharp decline in the living standard among the non-rubber growers. At present the state government tacitly favors rubber plantation at the cost of food security which can lead to a serious consequence on social and economically vulnerable groups of people living in Tripura, Sharma, George K. and Dey (2014) analyzed the contributions of various institutional interventions in the growth of rubber plantation sector of Tripura. The analysis reveals that the success and development of rubber plantation in the state are mainly due to the collaborative approach between major production-centric institutional interventions viz. TFDPCL, TRPCL, Block Planting Units-Rubber Producers Societies (BPU-RPS) schemes of Rubber Board and the marketing interventions such as establishing of centrifuge factory by TFDPCL, entry of Manimalayar Trading Company promoted by Rubber Board and increased participation of private traders. They suggest that development of rubber based industrial sector in the state to ensure value addition fetching higher income and employment opportunities enhancing the economy of the state.

Focus area of study addressed in the present book

- It examines the spatial pattern of growth and distribution of area rubber plantation in the state.
- It assess the role of different agencies on the growth and distribution of rubber plantation in the state.

- It evaluate the contributions of rubber plantation on the socio-economic development of the marginalized ST families in providing them with alternative livelihood options.

Figure 1.1: Location map of Tripura.

Database and Method Used in the Study

Source of Secondary Data

Secondary data regarding area under rubber plantation, rubber production, district wise area under rubber plantation of the state has been collected from the Rubber Board, Agartala, Tripura.

Annual data regarding area under rubber plantation of the state as well as the district wise area under rubber plantation managed by the different agencies has been collected from the specific agencies *viz.* TRPCL, TFDPCL, and the Rubber Board, Agartala, Tripura.

Related data has also been collected from Tripura Forest department, Agriculture department of Government of Tripura and Department of Revenue (directorate of land records and settlement, Tripura), Tripura tribal area autonomous district council (TTAADC), Sub-divisional district magistrate (SDM) and Block development office (BDO).

Satellite data *viz.* Landsat 5 TM (2001) and Landsat 8 OLI/TIRS (2015) images has been acquired from United State Geological Survey (USGS) Earth Explorer.

Related data has also been collected from books, articles and websites etc which has been duly acknowledged.

Method used for Analyzing Secondary Data

Annual data regarding area under rubber plantation and district wise area under rubber plantation of the state maintained by different agencies has been collected in order to compute time series analysis suggesting the spread of rubber plantations in the state. This has been further corroborated with analysis of satellite imageries.

Satellite data of 2001 and 2015 are obtained in order to assess the spatial distribution of area under rubber plantation in the state.

Different statistical techniques used to analyze the expansion of area under rubber plantation, 'arithmetic rate of growth' has been applied to suggest the expansion of area under rubber plantation in the state.

The formula for which is as follows-

$r = (P_t - P_o)/ P_o \times 100$

where,

r = rate of growth,

Pt = the interval between base year and terminal year and Po = area at the base year.

In order to find out the decadal growth rate of area under rubber plantation, 'annual compound rate of growth' has been used.

The formula for which is as follows-

$P_t = P_o (1+ r)t$

where,

P_o = area at the base year

r = rate of growth and

t = the interval between base year and terminal year.

To evaluate the time series trend of annual gain of area under rubber plantation '3 yearly moving average has been applied. The formula is as follows

M.A. (3 yearly)=a+b+c/3, b+c+d/3, c+d+e/3, d+e+f/3, e+f+g/3..............

Here,

a,b,c,......are the values in the series.

To find out the concentration of area under rubber plantation in the different districts of the state, 'Concentration Index' (Hassan, 2005) has been used, the formula for which is as follows-

$CI= (P/\nabla P)/(A/\nabla A)$

where,

CI = Concentration Index,

P = Actual area under rubber plantation of the ith district,

∇P = Average area under rubber plantation of the district in the state,

A = Actual area size of the ith district, and

∇A = Average area size of the district in the state.

This has been further mapped by appropriate cartographic technique.

Mean, Standard deviation, Skewness and Correlation has been used in order to assess the average, variation and skewness of the annual gain of area under rubber plantation in the state. Correlation has also been computed in order to find out the relationship between area under rubber plantation and production of NR in the state.

The spatial distribution of area under rubber plantation in the state, has been assessed with the help of satellite images that gives access to past records which has been compared present satellite data. In addition, geographic information system (GIS) has been used as this is a powerful tool for analysis of the natural resources and its interpretation. 'Remote sensing is highly useful for updating and monitoring the dynamics of land use change through the multi-temporal satellite data' (Trisurat et al., 2000, cited by Meti, Pradeep, Jacob, Meerabai and Jessy, 2014). 'The utility of remote sensing and GIS technique in assessing the suitability of natural resources in similar issues has been proved by earlier studies' (Sharma et al., 1989, Rao, 1991, Dhinwa et al., 1992, Kumar et al., 2002, Navalgund, 2002, cited by Meti et al.,2014).

As defoliation of rubber trees takes place during the month of January-February (Raj et al. 2012). By the month of March the rubber trees are fully re-foliated. Consequently, satellite datasets for the month of March as master data have been used for estimating the spatial extent of area under rubber plantation in Tripura. The satellite imageries of the month of re-foliation/foliation phase of rubber trees has been selected for examining the spatial distribution of area under rubber plantation in the state.(table 1.1), Landsat 5 TM images of 19thMarch 2001 and Landsat 8 Operational Land Imager and Thermal Infrared Sensor (OLI/TIRS) images of 26th March 2015 has been extracted so as to assess the difference in the spatial distribution of are aunder rubber plantation in Tripura. The Land sat images acquired from Landsat 5 TM and Landsat 8 OLI/TIRS consists of two images each with the same Path/Row: 136/43 and 44 with 30 metres (m) resolution. These two satellite scenes each from Landsat 5 TM and Landsat 8 OLI/TIRS has been acquired as it covers the entire state of Tripura.

The dates of the Landsat images for Landsat 5 TM and Landsat 8 OLI/TIRS vary depending upon the availability of cloud free or acceptable cloud contamination images available in the foliation stage.

Table 1.1: Specifications of satellite image used

Satellite	Sensor ID	Path /Row	Resolution	Acquisition date
Landsat 5 TM	LT05_L1TP_136043_20010319] _20161212_01_T1	136/43	30 m	19-MAR- 01
	LT05_L1TP_136044_20010319 _20161212_01_T1	136/44	30 m	
Landsat 8 OLI/ TIRS	LC08_L1TP_136043_20150326 _20170411_01_T1	136/43	30 m	26-MAR- 15
	LC08_L1TP_136044_20150326 _20170411_01_T1	136/44	30 m	

Source: https://earthexplorer.usgs.gov.

ERDAS IMAGINE 9.1 and Arc GIS 10.3 software has been used. ERDAS IMAGINE has been used for image processing while Arc GIS has been used for analyzing the image.

Following are schematic of work flow for image classification:-

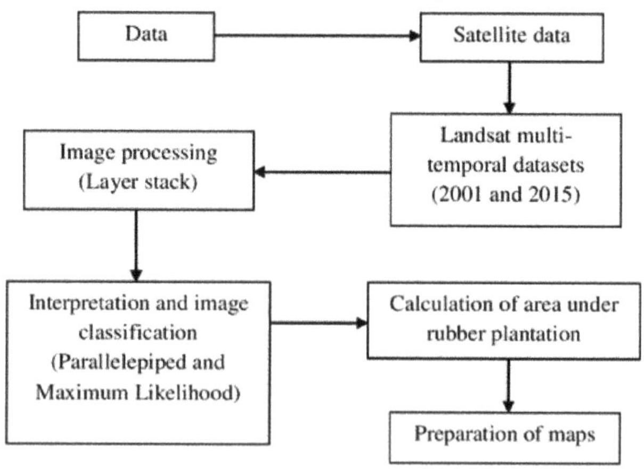

Image Pre-processing

The Landsat 5 TM images and Landsat 8 OLI/TIRS images has been pre-processed by the help of ego-referencing, layer stacking and mosaic in order to Or tho- rectify the images. Here, for each image of the Landsat 5 TM spectral bands from one to five and seven and for each image of the Landsat 8 OLI/TIRS spectral bands from one to seven and nine which is in the form of Tagged Image File Format

(TIFF) files are imported in the ERDAS IMAGINE software for layer stacking. After that the two- layer stack images each from Landsat 5 TM and Landsat 8 OLI/ TIRS are mosaic respectively, as already mentioned it covers the entire study area i.e. Tripura.

Image Processing

The mosaic images of Landsat 5 TM and Landsat 8 OLI/TIRS respectively are imported to ArcGIS software to extract the study area image and to classify the image. Here, the study area boundary is acquired from 1: 250,000 scale toposheet map (Army Map Service,1955).

The Landsat images are geo-rectified (Geometric Correction) to the Universal Transverse Mercator (UTM) coordinate system Zone 46 N and World Geodetic System (WGS 84) datum.

Image visualization enhancements and transformations are applied on the extracted Landsat images of the study area to assist the classification process for both the Landsat 5 TM and Landsat 8 OLI/TIRS. In order to improve the visualization of digital imagery the false colour composite with band combinations of Red (band 5), Green (band 4) and Blue (band 3) are applied in the Landsat 5 TM and band combinations of Red (band 6), Green (band 5) and Blue (band 4) are applied in the Landsat 8 OLI/TIRS. Applying the band combinations of R/G/B i.e. 5/4/3 and 6/5/4 for both the Landsat 5 TM and Landsat 8 OLI/TIRS respectively the imagery in the foliation duration, display rubber plantation appear light green that depicts and clearly distinguishes rubber plantation from others surrounding features. Dong, Xiao, Chen, Torbick, Jin, Zhang and Biradar (2013) also highlights that the false colour composite of (R/G/B = Band 5/4/3) during the foliation phase (March) display rubber plantations to appear light green and suggests that the images of rubber plantation formed unique phonological phases and have better performance in comparison to other periods.

Image Classification

The extracted Landsat images of Tripura , both the Landsat 5 TM and Landsat 8 OLI/TIRS are classified using a supervised classification procedure in ArcGIS software.

In supervised classification, the analysts trains the software (ArcGIS) to recognize the training samples in the image where pixels are selected that best represent the same patterns or land cover features of each category that the software recognizes. Here, a group of polygon regions of interest (RoIs) or Areas of interest (AoIs) are manually digitized on the Landsat images (Landsat 5 TM and Landsat 8 OLI/TIRS) and training samples are created.

Google Earth is used as reference for developing RoIs or AoIs. However, as the main motive is to examine the spatial distribution of rubber plantation therefore to ensure high classification accuracy and to prevent misclassification; a combination of distribution map of rubber plantation classified from satellite image (Raj et al., 2012) and Google Earth has been used as reference for developing RoIs or AoIs and training sample particularly for rubber plantation.

After the RoIs or AoIs are digitized and training samples developed, a signature file (SIG) is created. The SIG file contains a variety of information where each pixel categorized the features that mostly resemble each category. After that the SIG file created is selected and is put in the classification process of the Maximum Likelihood Classification for image classification.

After classification of the image others category features are ignored and only rubber plantation has been considered as the main purpose is to map and examine the spatial distribution of rubber plantation.

Accuracy Assessment

After image classification, accuracy assessment of the rubber plantation has been evaluated in regards to mapping accuracy. Google Earth has been used for accuracy assessment; as studies suggest that Google Earth produce high accuracy for Land use Land cover classification (Tilahun and Teferie, 2015; Sun, Leinenkugel, Guo, Huang and Kuenzer, 2017). Here, 50 Random Points has been generated from the polygon RoIs or AoIs digitized for rubber plantation category in the Arc GIS and the Random Points are converted to Keyhole Markup Language (KML) in order to open in the Google Earth. After opening the Random Points in Google Earth each Random Point's value has been verified for accuracy assessment so as to measure how many ground truth pixels area are correctly classified. Mapping accuracy of rubber plantation has been calculated using the formula (Raj et al., 2012) which is as follow:-

$$Accuracy\ (\%) = \frac{No.\ of\ correctly\ interpreted\ sample\ or\ reference\ random\ points}{Tqtal\ no\ ofsample\ or\ reference\ random\ points} \#\ 100$$

In addition, the report compiled by the Rubber Board on the total area under rubber plantation in the state has been consulted to estimate and verify with the classified area under rubber plantation in the mentioned Landsat imageries.

Rubber Plantation Map

The rubber plantation classified image is then extracted from the supervised classification image in order to identify the spatial pattern in the distribution of rubber plantation for the period of 2001 and 2015.

The rubber plantation map has been over layed on the land suitable for rubber plantation map and physiography map of the state of Tripura in order to assess the influence of land suitable and topography on the distribution of rubber plantation in the state of Tripura.

Source of Primary Data

Primary data has been collected through field survey in four selected rubber plantation centres (RPCs)/villages where TRPCL has developed rubber plantation. Here, RPC refers to a village where rubber plantation has been developed by different agencies. Primary data has been generated from the selected RPCs/ villages developed by the TRPCL. Information has been gathered at two levels-

1. Through focus group discussion, and

2. Household survey, where two sets have been selected i.e. first set comprise of beneficiaries households (BHHs) who are rehabilitated through rubber plantation scheme and practice rubber plantation and the second set comprise of households not practicing rubber plantation (NBHHs) but living in the same RPC/village.

Methodology for Generating Primary Data

Selection of District

In Tripura, at present there are eight districts *viz.* North Tripura, Unokoti, Dhalai, Khowai, West Tripura, Shepahijala, Gomati and South Tripura districts carved out from the previous four districts of North Tripura, Dhalai, West Tripura and South Tripura districts. The new districts are Unokoti (carved out from previous North Tripura district), Khowai and Shepahijala districts (carved out from previous West Tripura district) and Gomati district (carved out from the previous South Tripura district) has been created for administrative purpose by the government of Tripura on 21stJanuary, 2012 (DESPD, 2012). In this research, the previous four districts has been focused as the annual data of area under rubber plantation and number of RPC etc for the old districts are available and the new districts are recent and has already been included within the previous four districts.

Selection of Agency for the Study

As already mentioned, there are different agencies involved in promoting rubber plantation in the state as well as in rehabilitation programme of the marginalized section of people. In this research, the TRPCL has been selected, as the main objective of TRPCL is to economically rehabilitate the ST *jhumias*, landless and small farmers principally through rubber plantation. Besides this, TRPCL has its presence in all the four districts of the state as TRPCL managed RPCs are found in all the districts. Moreover TRPCL records the highest rehabilitated beneficiaries in comparison to the other agencies. TRPCL also has the highest number of RPCs than others agencies (table 1.2).

Table 1.2: RPC managed by the different agencies

Agency	No. of RPC
TFDPCL	67
TRPCL	68
RB (TBPP/BPS)	59

Source: TFDPCL, 2016a; TRPCL, 2016; Rubber Board, 2016.

Note 1: Other agencies like TTAADC and SDM/BDO are excluded from the above table as their contribution are less.

Note 2: Out of 67 RPCs managed by TFDPCL, only 29 centres are for resettlement of jhumias, landless and other marginalized section belonging to ST and Scheduled Castes (SC) population while 38 RPCs are for commercial purpose.

Selection of Field Study Site

For selection of field study site, firstly, the 68 RPCs managed by the TRPCL distributed in all the four districts have been arranged as per district, which has been further categorized block wise. Secondly, the blocks having maximum RPC in each district has been identified, and then one RPC has been selected randomly from these blocks representing each district for field investigation. The four selected RPCs/villages for field investigation are Depachhara RPC located in Pencharthal block of North Tripura district, Dakshin Dhumachhara RPC located in Manu block of Dhalai district, Pathaliaghat RPC located in Bishalgarh block of West Tripura district and Dakshin Taidu RPC located in Ampi block of South Tripura district (table 1.3). Location map of the four selected RPCs/villages for field investigation is shown in figure1.2.

Table 1.3: RPC managed by the TRPCL in all the district of Tripura

Sl. No.	Name of Centre	Block & no. of centre	Centre/village selected for field survey
North Zone (North Tripura District)			
1	PurbaDalubari	Panisagar (1 centre)	
2	Chinibagan	Gournagar (2 centre)	
3	Dhanbilash		
4	Sadhugaon	Kumarghat (3 centres)	
5	Fatikchhara		
6	Betchhara		
7	Damchhara	Damchhara (1 centre)	
8	Dasamani Para	Dasda (3 centres)	
9	Kangraichhara		
10	Jayantipur		
11	Thalibari	Pencharthal (6 centres)	Depachhara
12	Nabinchhara		
13	Depachhara		
14	Jainhumpara		
15	Panbagan		
16	Sailenbari		
Dhalai Zone (Dhalai District)			
1	Abhanga	Salema (6 centres)	
2	Katalutma		
3	Srirampur		
4	Bilashchhara		
5	Mechuria		
6	Kachuchhara		
7	Lalcheri-Bagmara	Ambassa (2 centres)	
8	Kachimchhara		

Sl. No.	Name of Centre	Block & no. of centre	Centre/village selected for field survey
9	Lambabil-Mainama	Manu (7 centres)	Dakshin Dhumachhara
10	Kathalchhara		
11	Dakshin Dhumachhara		
12	East Karamchhara		
13	Kanchanchhaa		
14	Naitangchhara		
15	Ultachhara		
16	Gandachhara RPC	Dumburnagar(1 centre)	
West Zone (West Tripura District)			
1	Takchaya	Tulashikhar (2 centres)	
2	Badlabari		
3	Akhrabari	Padmabil (1 centre)	
4	Lefunga	Mohanpur (2 centre)	
5	S. Nagar TRPC project		
6	Belbari	Jirania (1 centre)	
7	Surendranagar	Hezamara (1 centre)	
8	Uttar Promodenagar	Mungiakami (1 centre)	
9	NilpurnaKalaiPara	Dukli (1 centre)	
10	Paschim Takarjala	Jampuijala (3 centres)	
11	Kandraichhara		
12	Ghaniamara		
13	Herma	Bishalgarh (4 centres)	Pathaliaghat
14	Sutramura		
15	Pathaliaghat		
16	Bangshibari		
17	Manaipathar	Kathalia (1 centre)	
18	Taibandal	Melaghar (1 centre)	
19	Bijoynagar	Boxanagar (1 centre)	
South Zone (South Tripura District)			
1	Purba-PaschiDaluma	Amarpur (1 centre)	
2	Sonachhara	Ampi (3 centres)	Dakshin Taidu
3	Chechua		
4	Dakshin Taidu		
5	Karbook RPC	Karbuk (1 centre)	
6	Lovestory	Matarbari (1 centre)	
7	Laxmipati-Photamati	Killa (3 centres)	
8	Twicharbari		
9	Choygharia		
10	Dhuptali	Kakarban (1 centre)	
11	Gaurifa-Chalitacherri	Rupaichhari (1 centre)	

Sl. No.	Name of Centre	Block & no. of centre	Centre/village selected for field survey
12	Bhuratali	Satchand (3 centres)	
13	Tuikumbachhara		
14	Gardhang		
15	East Pillak	Bokafa (3 centres)	
16	Salthangmanu		
17	Karaicherra		

Source: TRPCL, 2016.

Selection of Households and Sampling Technique

In this research, as already mentioned two sets of households have been selected for conducting household survey viz. Beneficiary households (BHHs) and non Beneficiary households (NBHHs) but belonging to the same community and living in close proximity to each other in the selected RPC/villages. The BHHs have been selected in order to analyze the impact of rubber plantation-based rehabilitation programme of the TRPCL on the socio-economic impact of the programme on the beneficiaries. Both the BHHs and NBHHs have been selected in order to compare the socio- economic conditions of the BHHs with that of the NBHHs, in order to assess how far this rehabilitation programme has impacted in enhancing the socio-economic conditions and diversification of livelihood practices of the BHHs and NBHHs.

For surveying the BHHs and NBHHs, stratified sampling technique has been used in the selection of the households from each selected RPC/village. Consequently stratified random sampling technique has been used in selection and surveying of the two sets of house holds.

The sample survey used is both descriptive and analytical in nature. It is descriptive as in descriptive survey certain information are collected from the populations' i.e. both BHHs and NBHHs, and it is analytical as in analytical survey comparisons are made between different groups i.e. BHHs and NBHHs. 'Many surveys provide data that serve both the purposes' (Cochran, 1977).

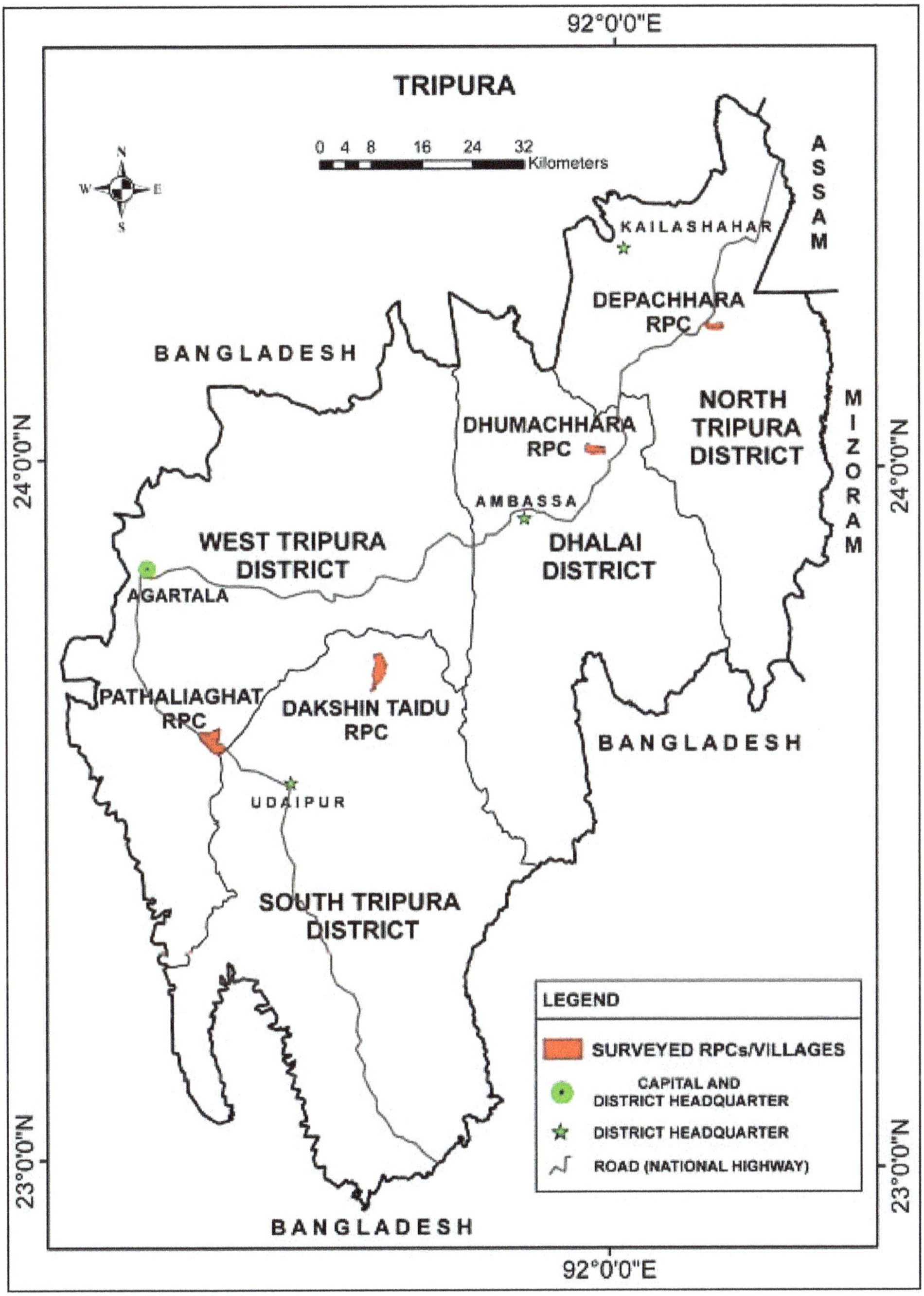

Figure 1.2: Location map of the surveyed RPCs/villages.

Sampling size

To determine appropriate sample size, three criteria are important i.e. the level of precision, the level of confidence or risk and the degree of variability in the attributes being measured (Miaoulis and Michener, 1976, cited by Israel, 1992).

The sample size for gathering primary data has been calculated using Cochran formula (Cochran, 1963, cited in Israel, 1992). The formula is as follow:-

1. $n = n_o/1+ ((n_o-1)/N)$

Where, n is the sample size, N is the population size, nois the required return sample size. Here, nois acquired applying the formula which is as follow:-

2. $n_o = Z^2pq/e^2$

Where, nois the sample size, Z^2 is the value for the selected alpha level i.e. confidence level, p is the estimated proportion of an attribute that is present in the population, q is 1-p and e is the desired level of precision/acceptable margin of error. The formula (2) has been developed to yield a representative sample for proportions from the population at large, but as variability in the proportion is not known, therefore, p=.5 (maximum variability) is taken. Further to adjust and reduce the necessary sample size for small populations, the formula (1) called the finite population correction for proportions has been applied, as the given sample size provides proportionately more information for a small population than for a large population.

The empirical data in this book has 95 percent confidence level i.e. 1.96 and precision/acceptable margin of error of ±4 percent i.e. .04.

To determine the sample size i.e. finite population, formula (2) has been applied which is as follow:-

$n_o = (1.96)2(.5)(.5)/(.04)2 = 600.25$

The sample size (no) i.e. 600.25 is a finite population at 95 percent confidence level and ±4 percent precision.

To determine the sample size from a total of 318 BHHs, formula (1) has been applied which is as follow:

n = 600.25/(1+ ((600.25-1)/318)) = 208 BHHs

Therefore, the required sample size is 208 BHHs out of the total 318 BHHs living in the four selected RPCs/villages. To survey 208 BHHs, from each RPC/ village about 50 percent of the BHHs which has been rounded to 210 BHHs that has been surveyed. Table 1.4, indicates that there are 318 BHHs living in the four selected RPCs/villages. From these 318 BHHs, 210 households were surveyed using simple random technique for collection of data as suggested by the table 1.4

Table 1.4: RPC selected for field study and number of household surveyed

District	Center/Village selected for field study	No. of BHHs	No. of BHHs surveyed	Surveyed BHHs to total (in %)
North Tripura	Depachhara	76	54	71
Dhalai	Dakshin Dhumachhara	84	45	54
West Tripura	Pathaliaghat	72	56	78
South Tripura	Dakshin Taidu	86	55	64
Total	4 RPCs	318	210	

Source: Field survey, 2016.

For the second set of households *i.e.* NBHHs, about 101 households have been surveyed using simple random sampling technique *i.e.* 24 NBHHs from Depachhara, 26 NBHHs from Dakshin Dhumachhara, 23 NBHHs from Pathaliaghat and 28 NBHHs from Dakshin Taidu are surveyed for collection of data.

The household survey was conducted during the month of February to July, 2016.

Data Collection

Information regarding the general background of the four RPCs/villages before and after the rubber plantation-based rehabilitation programme has been introduced were collected through focus group discussion (FGD) with the village elders and the TRPCL official within the selected RPCs/villages.

Data on the socio-economic characteristics of the household, livelihood practices, indicators suggesting livelihood assets consisting of five assets/capitals *viz.* Human, Natural, Physical, Financial and Social assets/capitals has been collected from both the BHHs and NBHHs with the help of semi-structured schedule and by interviewing the head of the house holds.

Data regarding livelihood practices directly related to rubber plantation and livelihood practices not related with rubber plantation from both the BHHs and NBHHs has been collected

Data regarding the livelihood assets has been collected from both the BHHs and NBHHs so as to analyze the impact of rubber plantation on their socio-economic conditions. The sustainable livelihood framework has been used as it is an analytical tool developed by the Department for International Development (DFID), in 1999 to measure the various dimension of livelihood of the socio-economically deprived population. This people-centered framework has been used to demonstrate the main factors affecting people's livelihoods, and also depicts the typical relationships between the assets. This framework helps to understand and analyze the livelihood conditions of the poor and is a useful tool in guiding how to reduce poverty level. This framework also underlines the significance of linkages between household assets and their effect on livelihood in diverse socio-economic contexts. As the sustainable livelihood framework is versatile and flexible tool, it has been used to assess the socio-economic development associated with rubber plantation. The five livelihood assets/capitals i.e. human, social, natural, physical and financial assets/capitals mentioned in the framework have been computed by information collected through suitable indicators from the surveyed households.

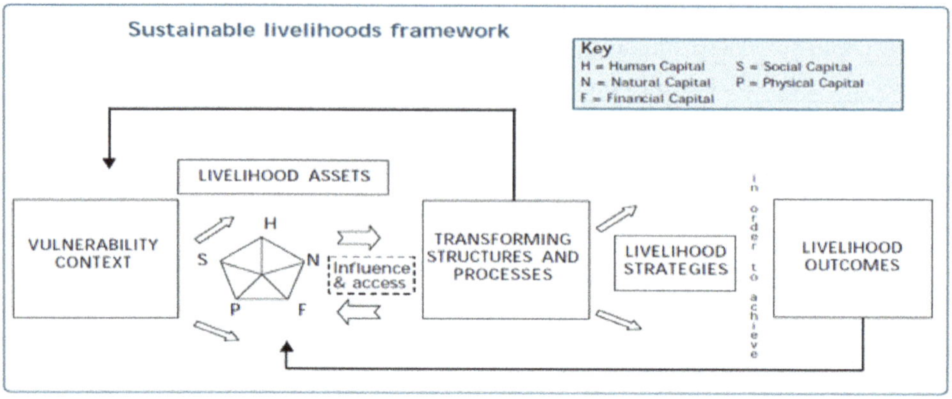

Figure 1.3: Sustainable livelihoods framework of the DFID.

Source: DFID, 1999.

Concept of Sustainable Livelihood

'A livelihood is sustainable when it can cope with and recover from stresses and shocks and maintain or enhance its capabilities and assets both now and in the future, while not undermining the natural resource base' (Chambers and Conway, 1992, cited by DFID, 1999). 'Poverty analyses have shown that people's ability to escape from poverty is critically dependent upon their access to assets' (DFID, 1999). As per Carney (1998, cited by Krantz, 2001) the sustainable livelihood framework is built around five livelihood assets/capitals, graphically depicted as a pentagon to underline their interconnections and the fact that livelihoods depend on a combination of assets/capitals of various kinds and not just from one category. An important part of the analysis is thus to find out people's access to different types of assets/capitals (physical, human, financial, natural and social) and their ability to put these to productive use.

The indicators for computing the five assets/capitals for assessing how sustainable is rubber plantation as a livelihood option for the households practicing rubber plantation are as follows.

Indicators for human asset/capital: (i) working population, (ii) literacy rate, (iii) traditional skills known, and (iv) modern skills known.

1. **Indicators for natural asset/capital:** (i) size of landholding, (ii) access to land for *jhum* cultivation, (iii) access to common water resources, and (iv) access to common forest-resources.
2. **Indicators for physical asset/capital:** (i) access to rubber processing center within the PRC/village, (ii) access to blacktop road within a kilometer (km), (iii) access to market within 5 km, (iv) access to safe drinking water

sources within a meter from the house, (v) access to school within the RPC/village, and (vi) access to electricity in the house.

3. **Indicators for social asset/capital:** (i) good relationship with TRPCL officials of the RPC/village, (ii) training and capacity development activities, (iii) committee members of the locality, (iv) member of self-help groups, and (v) participation in programme or meeting organized by rubber growing agencies and others government and non-government organization regarding rubber and others.
4. Indicators for financial asset/capital are mainly focused on income from rubber plantation and other sources which has been categorized into in come groups.

Analysis of Primary Data

Livelihood Diversification

There are different measures of livelihood diversification, some of this measurement indices are- number of income sources and their share, Simpson index, Herfindahl index, Ogive index, Entropy index, Modified Entropy index and Composite Entropy index (Shiyani and Pandya, 1998, cited by Khatun and Roy, 2012).

In this research, to measure and compare the livelihood diversity of both the BHHs and NBHHs, a modified version of Simpson Index of Diversity (SID) (Simpson, 1949) has been used.

Simpson (1949) was the first who proposed index for measuring diversity and concentration of population species in ecology. The index proposed by Simpson was further modified to avoid difficulty of heterogeneity (Green bery, 1956 and Berger and Parker, 1970, cited by Peet, 1974) proposed subtracting the Simpson index from its maximum possible value of 1; this formulation was originally suggested by Gini, 1912 and Pielou, 1969 (cited by Peet, 1974). The modified formula of Simpson diversity index are:-

$D=1 -\sum p^2$ (Gini Simpson index formula) and

$D = 1-\Sigma \{[n_i(n_i-1)]/[N(N-1)]\}$ (Pielou Simpson index formula).

Applying the modified formula of Simpson index of diversity (SID) several researchers (Saha and Bahal, 2010 and 2014; Fabusoro et al., 2010; Khatun and Roy, 2012; Agyeman and Onumah, 2014; Anandan et al., 2014; Ahmed et al., 2015; Debele and Desta, 2016; Dev, Sultana and Hossain, 2016; Addisu, 2017) has measured livelihood diversification. Khatun and Roy (2012) have used 'SID because of its computational simplicity, robustness, and wider applicability'. Thus, in this research, the modified formula of SID has been used for measuring the diversification of livelihood practices of both the BHHs and NBHHs. SID is expressed by the formula -

$SID = 1-\Sigma \{[ni(ni-1)]/[N (N-1)]\}$

Where, N is the total numbers of livelihood practice/earning sources and nire presents the proportion of livelihood practice of ith livelihood practice. SID value ranges between 0 and 1. In case if there is just one source of livelihood practices ni = 1 than SID is 0. The closer the value of SID is to zero, more is the specialization, and the further it is from 0 and closer to 1, it implies more diversification.

Five capitals of the livelihood assets

To evaluate the five assets/capitals of the households (BHHs and NBHHs), Range Equalization Method (UNDP method of human development index- (Ghosh, Kumar, Roy and Bairy, 2006) has been used so to compute indices for each indicator and to aggregate them into composite index score for each of the five assets/capitals. Firstly, the values of every indicator of the five assets/capitals are converted into percentage values. Secondly, the value of each indicator of the five assets/capitals is measured in order to obtain scale-free aggregate indices value ranging between 0 and 1, with higher value suggesting greater strength. The normalized scale-free aggregate indices value ranging from 0 to 1 for each indicator of the five assets/capitals are obtained by subtracting the minimum percentage value from each indicator and then dividing it by the indicator-range i.e. maximum percentage value subtracted by minimum percentage value. The minimum and maximum values are set in order to transform the indicators expressed in different units into indices between 0 and 1. The formula for obtaining scale-free value is asfollows:-

$$Svi = (X\text{-}Min) / (Max\text{-}Min)$$

where,

Svi = scale free value (ranging 0 to 1) of the indicator in the ith capital, X = actual value of the indicator in the ith capital,

Min = minimum value of the indicator in the ith capital, and Max = maximum value of the indicator in the ith capital.

Here, the scale-free values are obtained so that while using composite index there is no bias toward the highest indicator values.

Thirdly, the scale free values of each indicator obtained for each asset/capital respectively are added and divided by the total strength/number of indicators of each asset/capital respectively to obtain the composite index score for each asset/ capital. The composite index score ranges from 0 to 1. The formula for obtaining composite index score is as follows:-

$$CIS = \sum Svi / N$$

where,

CIS = composite index score,

Svi = total scale free value of the ith capital, and N = total number of indicator in the ith capital.

Applying the formula, composite index score of each of the five assets/capitals of the livelihood assets are obtained.

To determine livelihood sustainability as per the score of composite index; ranging from 0 to 1 has been compared to the theoretical norm where the composite index score is categorized into four groups suggesting the level of sustainable livelihood. The theoretical norm of the composite index score are: 0 - 0.25 (low sustainability), 0.26 - 0.50 (moderately sustainable), 0.51 - 0.75 (moderately highly sustainable) and 0.76 - 1 suggestive of highly sustainable (Kamaruddin and Samsudin, 2014). The composite index score derived for the five assets/capitals of both the BHHs and NRHHs has been diagrammatically represented by livelihood asset pentagon so to visualize the strength and weakness of each asset/capital of both the BHHs and NBHHs livelihood assets. "The asset pentagon lies at the core of the livelihoods framework, 'within' the vulnerability context" (DFID,1999).

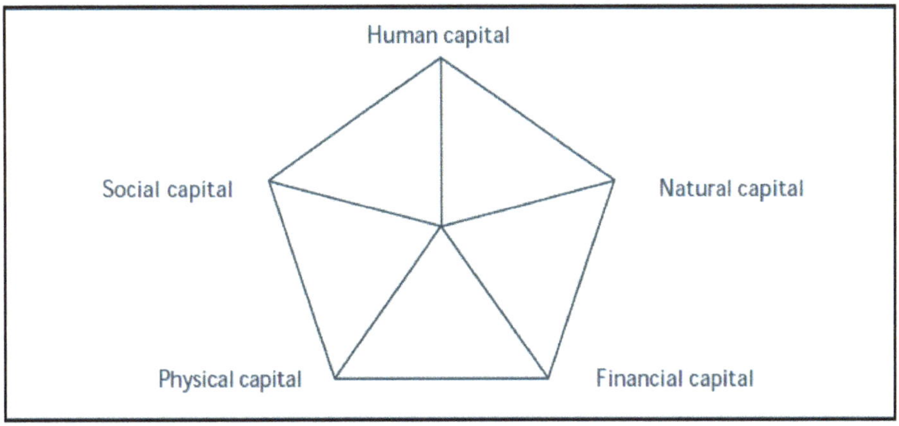

Figure 1.4: Livelihood asset pentagon.

Source: DFID, 1999.

Note: The centre point of the pentagon represents zero access to assets while the outer perimeter represents maximum access to assets/capitals. On this basis, variation in access to assets can be drawn in different shaped pentagons of the five assets/capitals.

Limitations

Some of the limitations are

1. Though rubber plantation in the state was introduced in the year 1963, continues data series of area under rubber plantation is not available. It is from 1972 onwards that continues data of area under rubber plantation is available, consequently the time series analysis, nature of gain, annual and decadal growth rate of area under rubber for the state are analyzed and discussed from 1972 to 2016.
2. The district wise area under rubber plantation are available from 2008 onwards, consequently district wise distribution of area under rubber plantation could not be analyzed prior to 2008.
3. At present Tripura has eight districts which has been created in 2012 from the previous four districts, consequently the time series analysis, growth rate of area under rubber plantation etc has been computed for the previous four districts (which include the administrative boundary of the newly formed districts).

4. The list of the RPCs/ villages of TRPCL developed rubber plantation for the present eight district of Tripura are not available, consequently the previous four districts of the state have been selected for household survey.

Design of the Book

The book is divided into five chapters, the first chapter is introductory in nature highlighting the issues this book addresses, data source and method adopted to realise the objectives of the book are discussed in detail.

The second chapter basically deals with the geographical personality of Tripura along with a short history of the erstwhile kingdom of Tripura, a princely state during the colonial era.

The focus of the third chapter is on the growth and distribution of rubber plantation in the state.

The fourth chapter deals with the role of different agencies in the distribution and growth of rubber plantations.

The fifth chapter analyses the socio-economic benefits associated with rubber plantations. It also highlights on livelihood diversification among the beneficiary households which has been compared to the socio-economic condition of the households living in the same villages but not covered by the rubber rehabilitation scheme.

References

Addisu, Y. 2017. Livelihood strategies and diversification in western tip pastoral areas of Ethiopia, Pastoralism: Research, Policy and Practice, 7:9, pp.1-9.

Agyeman, B.A.S., Brempong, S.A. and Onumah, E.E. 2014. Determinants of Income Diversification of Farm Households in the Western Region of Ghana, Quarterly Journal of International Agriculture, Vol. 53, No. 1, pp.55-72.

Ahmed, M.T., Bhandari, H., Gordoncillo, P.U., Quicoy, C.B. and Carnaje, G.P. 2015. Diversification of rural livelihoods in Bangladesh, Journal of Agricultural Economics and Rural Development, Vol. 2(2), pp.032-038.

Ahrends, A., Hollingsworth, P.M., Ziegler, A.D., Fox, J.M., Chen, H., Su, Y. and Xu,

Anandan, G., Thomas, A., Benickson, C., Chitra, DR., Geethu. M., et al. 2014. Estimation of Tree Species Diversity in Four Campuses of Roever Institutions using Simpsons Diversity Index, Journal of Biodivers Endanger Species Volume 2, Issue 4, pp. 135.

Army Map Service. 1955. Toposheet No. NF 46-1, NF 46-2, NF 46-6, NG 46-13 and NG 46-24, Series U502, Scale 1: 250,000, Corps of Engineers, U.S.Army, Washington, D.C.

Bareh, H.M. 2001. Encyclopaedia of North-East India-Volume viii-Tripura, Mittal Publications, New Delhi, pp. 1-279.

Bareh, H.M. 2001. Encyclopaedia of North-East India-Volume viii-Tripura, Mittal Publications, New Delhi, pp. 1-279.

Bhattacharjee, S.R. 1989. Tribal Insurgency in Tripura: A Study in Exploration Causes, Inter- India Publications, New Delhi, pp. 23-28.

Bhattacharya, B.K. 1980. Pattern of Administrative Organisation, in J.C. Chaudhuri (ed.) Tripura- The Land and its People, Leeladevi Publication, Delhi, pp. 44- 54.

Bhattacharya, S. 1992. From Jhuming to Tapping, Directorate of Research, Department of Welfare for Schedule tribes, Government of Tripura, Agartala, pp. 1-85.

Bhowmik, I. 2006. A status on Rubber Plantations in Tripura, Tripura Rubber Mission- Technical Bulletin, Tripura, pp. 1-21.

Bhowmik, I. and Viswanathan, P.K. 2015. Emerging Labour Relations in the Small Rubber Plantations of Tripura, NRPPD, Discussion Paper No. 47, Centre For Development Studies, Thiruvananthapuram, pp. 1-33.

Bordaloi, B.N. 1990. The Problem of Rehabilitation of the Jhumiya Families in the Hills Areas of Assam, in D.N. Majumdar (ed.) Shifting Cultivation in Northeast India, Omsons Publication, New Delhi, pp.226-228.

Borthakur, D.N. 1992. Agriculture of the North eastern region: with special references to hill Agriculture, BeeceePrakashan, Guwahati, pp.59-71.

Burma, K.K. 1990. Shifting Cultivation in the Hills of Assam, in D.N. Majumdar, (ed.) Shifting Cultivation in Northeast India, Omsons Publication, New Delhi, pp. 226-228.

Chakravarti, M. 1986. Forestry in Tripura: Its Past and Present, in M.D. Gupta, A.K. Gangopadhyay, T. Bhattacharya, M. Chakraborti (ed.) Forestry Development in North-East India, Omsons Publications, Guwahati and New Delhi, pp. 31- 41.

Chakravarti, M. 1986. Forestry in Tripura: Its Past and Present, in M.D. Gupta, A.K. Gangopadhyay, T. Bhattacharya, M. Chakraborti (ed.) Forestry Development in North-East India, Omsons Publications, Guwahati and New Delhi, pp. 31- 41.

Chatterjee, S.N. 1984. Tripura- A Profile, Inter-India Publications, New Delhi, pp. 1- 47.

Chaudhuri, N.R.R. 1980. The Historical Past, in J.C. Chaudhuri (ed.) Tripura- The Land and its People, Leeladevi Publication, Delhi, pp. 13-43.

Chaudhuri, P.S., Bhattacharjee, S., Dey, A., Chattopadhyay, S., and Bhattacharya, D. 2013. Impact of age of rubber (Hevea brasiliensis) plantation on earthworm communities of West Tripura (India), Journal of Environmental Biology, Vol. 34, pp.59-65.

Chen, H., Yi, Z-F., Schmidt-Vogt, D., Ahrends, A., Beckschafer, P., Kleinn, C., Ranjitkar S. and Xu, J. 2016. Pushing the Limits: The Pattern and Dynamics of Rubber Monoculture Expansion in Xishuangbanna, SW China, PloS ONE 11(2), pp.1-15.

Choudhury, J. 2009. Efforts of Tribal Development in Tripura: An Assessment, , in

Cochran, W. G. 1977. Sampling techniques (3rd ed.), New York: John Wiley & Sons.

Das, S., Choudhury, S. and Roy, A. 2012. The Success Story of Rehabilitation of Jhumias in Tripura- A study on Baramura-Deutamura Range, International Journal of Engineering and Science, Vol. 1, Issue 10, pp. 25-29.

Das, S., Choudhury, S. and Roy, A. 2012. The Success Story of Rehabilitation of Jhumias in Tripura- A study on Baramura-Deutamura Range, International Journal of Engineering and Science, Vol. 1, Issue 10, pp. 25-29.

Dasgupta, M. 1986. Jhumias of Tripura, Economic and Political Weekly, Vol. 21, No. 44/45, pp. 1955-1957, 1959-1960.

Dasgupta, M. 1986. Jhumias of Tripura, Economic and Political Weekly, Vol. 21, No. 44/45, pp. 1955-1957, 1959-1960.

Dasgupta, M. 1986. Jhumias of Tripura, Economic and Political Weekly, Vol. 21, No. 44/45, pp. 1955-1957, 1959-1960.

Debbarma, R. 2013. Socio-Economic Impact Of Rubber Plantation On Shifting Cultivators In Dhalai District Of Tripura, M.Phil Dissertation, North-Eastern Hill University, Shillong, Meghalaya, pp. 54-135.

Debbarma, S. 2009. Refugee Rehabilitation and Land Alienation in Tripura, in W. Fernandes and S. Barbora (ed.) Land, People and Politics: Contest over Tribal Land in Northeast India, North-Eastern Social Research Centre, Guwahati and International Work Group for Indigenous Affiars (IWGIA), Denmark, pp. 113-127.

Debbarma, S. 2009. Refugee Rehabilitation and Land Alienation in Tripura, in W. Fernandes and S. Barbora (ed.) Land, People and Politics: Contest over Tribal Land in Northeast India, North-Eastern Social Research Centre, Guwahati and International Work Group for Indigenous Affiars (IWGIA), Denmark, pp. 113- 127.

Debbarma, S. 2009. Refugee Rehabilitation and Land Alienation in Tripura, in W. Fernandes and S. Barbora (ed.) Land, People and Politics: Contest over Tribal Land in Northeast India, North-Eastern Social Research Centre, Guwahati and International Work Group for Indigenous Affiars (IWGIA), Denmark, pp. 113- 127.

Debbarma, S. 2009. Refugee Rehabilitation and Land Alienation in Tripura, in W. Fernandes and S. Barbora (ed.) Land, People and Politics: Contest over Tribal Land in Northeast India, North-Eastern Social Research Centre, Guwahati and International Work Group for Indigenous Affiars (IWGIA), Denmark, pp. 113- 127.

Debbarma, S. 2009. Refugee Rehabilitation and Land Alienation in Tripura, in W. Fernandes and S. Barbora (ed.) Land, People and Politics: Contest over Tribal Land in Northeast India, North-Eastern Social Research Centre, Guwahati and International Work Group for Indigenous Affiars (IWGIA), Denmark, pp. 113- 127.

Debbarma, S. 2009. Refugee Rehabilitation and Land Alienation in Tripura, in W. Fernandes and S. Barbora (ed.) Land, People and Politics: Contest over Tribal Land in Northeast India, North-Eastern Social Research Centre, Guwahati and International Work Group for Indigenous Affiars (IWGIA), Denmark, pp. 113- 127.

Debele, B.N. and Desta, G.D. 2016. Livelihood Diversification: Strategy, Determinants and Chaallenges for Pastoral and Agro-Pastoral Communities of Bale Zone, Ethiopia, International Review of Social Sciences and Humanities, Vol. 11, No. 2, pp.37-51.

Department For International Development (DFID). 1999. Framework, Sustainable Livelihoods Guidance Sheets. Retrieved from www.livelihoods.org. Accessed on 23 November 2014.

DESPD. 2012. Economic Review of Tripura (2011-12), 13thissue, Government of Tripura, Agartala, Tripura, pp. 3-46.

DESPD. 2015. Economic Review of Tripura (2014-15), 16thissue, Government of Tripura, Agartala, Tripura, pp. 144-157.

Dev Varman, S.B.K. 1999. A Study Over The Jhum And Jhumia Rehabilitation In The Union Territory Of Tripura, Directorate of Research, Departmentof

Dev Varman, S.B.K. 1999. A Study Over The Jhum And Jhumia Rehabilitation In The Union Territory Of Tripura, Directorate of Research, Department of Welfare for Scheduled Tribes and Scheduled Castes, Government of Tripura, Agartala, Print Best, Agartala, Tripura, pp.9-39.

Dev, T., Sultana, N. and Hossain, M.E. 2016. Analysis of the Impact of Income Diversification Strategies on Food Security Status of Rural Households in Bangladesh: A Case Study of Rajshahi District, American Journal of Theoretical and Applied Business, Vol. 2, No. 4, pp. 46-56.

Devi, A.I. 2010. Amazing North East - Tripura, Vij Books India Pvt. Ltd. New Delhi, pp. 75-77.

Devi, A.I. 2010. Amazing North East - Tripura, Vij Books India Pvt. Ltd. New Delhi, pp. 75-77.

Devi, A.I. 2010. Amazing North East - Tripura, Vij Books India Pvt. Ltd. New Delhi, pp. 75-77.

Devi, A.I. 2010. Amazing North East - Tripura, Vij Books India Pvt. Ltd. New Delhi, pp. 75-77.

Devi, A.I. 2010. Amazing North East - Tripura, Vij Books India Pvt. Ltd. New Delhi, pp. 75-77.

DFID. 1999. Framework, Sustainable Livelihoods Guidance Sheets. Retrieved from www.livelihoods.org. Accessed on 23 November 2014.

Dong J. W., Xiao X. M., Chen B. Q., Torbick, N., Jin, C., Zhang, G. and Biradar, C. 2013. Mapping deciduous rubber plantations through integration of PALSAR and multi-temporal Landsat imagery, Remote Sensing of Environment, vol. 134, pp. 392-402.

Fabusoro, E., Omotayo, A.M., Apantaku, S.O. and Okuneye, P.A. 2010. Forms and Determinants of Rural Livelihoods Diversification in Ogun State, Nigeria, Journal of Sustainable Agriculture, 34:4, pp.417-438.

FAO. 2013. Food and Agriculture Organization of the United Nations, Processing of Natural Rubber. Retrieved from http://ecoport.org/ep?SearchType=earticleView&earticleId=644&page=4343. Accessed on 4 February 2016 and on 10thMarch 2017.

FAO. 2013. Food and Agriculture Organization of the United Nations, Processing of Natural Rubber. Retrieved from http://ecoport.org/ep?SearchType=earticleView&earticleId=644&page=4343. Accessed on 4 February 2016 and on 10thMarch 2017.

Fernandes and S. Barbora (ed.) Land, People and Politics: Contest over Tribal Land in Northeast India, North-Eastern Social Research Centre, Guwahati and International Work Group for Indigenous Affiars (IWGIA), Denmark, pp. 113-127.

Fox, J. and Castella, J.C. 2013. Expansion of rubber (Hevea brasiliensis) in Mainland Southeast Asia: What are the prospects for smallholders?,Journal of Peasant Studies, 40 (1), pp. 155-170.

Fu, Y., Brookfield, H., Guo, H., Chen, J., Chen, A. and Cui, J. 2009. Smallholder rubber plantation expansion and its impact on local livelihoods, land use and agrobiodiversity, a case study from Daka, Xishuangbanna, southwestern China, International Journal of Sustainable Development and World Ecology, Vol. 16, No.1, pp.22-29.

G.K. Bera, B.K. Medhi, R.P. Athparia, and K. Jose (ed.) Tribal Development in Tripura, EBH Publisher, Guwahati, pp.121-135.

George K. T., Haridasan, V. and Sreekumar, B. 1988. Role of Government and Structural Changes in Rubber Plantation Industry, Economic and Political Weekly,Vol. 23, No. 48, pp. M158-M166.

Geosrge, J.G. and Chandrashekar, H.M. 2014. Growth and Trends in Production and Marketing of Natural Rubber in Kerala, India, International Journal of Current Research and Academic Review, 2(8), pp.53-61.

Ghosh, S., Kumar, R., Roy, C.I. and Bairy, R. 2006. Social Development Index for Social Development, in A. Kundu (ed.) India Social Development Report, Council for Social Development, Oxford University Press, New Delhi, pp. 219-220.

Gupta, A.K 2000. Shifting Cultivation and Conservation of Biological Diversity in Tripura, Northeast India, Human Ecology, Vol. 28, No. 4, pp. 605-629.

Gupta, A.K. 2000. Shifting Cultivation and Conservation of Biological Diversity in Tripura, Northeast India, Human Ecology, Vol. 28, No. 4, pp. 605-629.

Gupta, A.K. 2000. Shifting Cultivation and Conservation of Biological Diversity in Tripura, Northeast India, Human Ecology, Vol. 28, No. 4, pp. 605-629.

Gupta, A.K. 2000. Shifting Cultivation and Conservation of Biological Diversity in Tripura, Northeast India, Human Ecology, Vol. 28, No. 4, pp. 605-629.

Gupta, A.K. 2000. Shifting Cultivation and Conservation of Biological Diversity in Tripura, Northeast India, Human Ecology, Vol. 28, No. 4, pp. 605-629.

Gupta, A.K. 2000. Shifting Cultivation and Conservation of Biological Diversity in Tripura, Northeast India, Human Ecology, Vol. 28, No. 4, pp. 605-629.

Gupta, A.K. 2000. Shifting Cultivation and Conservation of Biological Diversity in Tripura, Northeast India, Human Ecology, Vol. 28, No. 4, pp. 605-629.

Gupta, A.K. 2000. Shifting Cultivation and Conservation of Biological Diversity in Tripura, Northeast India, Human Ecology, Vol. 28, No. 4, pp. 605-629.

Gupta, B.N. 1990a. Statistics (Theory & Practice), Sahitya Bhawan, Agra, pp. 312-313, 515-584.

Gupta, B.N. 1990a. Statistics (Theory & Practice), Sahitya Bhawan, Agra, pp. 312-313, 515-584.

Gupta, M.D. 1990. Rehabilitating JhumiasOn Rubber Plantations: An Assessment of the Tripura Experiment, in D.N. Majumdar (ed.) Shifting Cultivation in North- East India, Omsons Publications, New Delhi, pp. 308-311.

Gupta, M.D. 1994. Continuity and Change in Tribal Societies: A Study of the Economic Aspects of Tribal Jhumia Society in Tripura, in M.S. Sangma (ed.) Essays On North-East India, Indus Publishing Company, New Delhi, pp. 145- 164.

Hasan, M.I. 2005. Population Geography, Rawat Publications, Jaipur, New Delhi, Bangalore, Hyderabad, Guwahati, Kolkata, pp.40-73.

Hasan, M.I. 2005. Population Geography, Rawat Publications, Jaipur, New Delhi, Bangalore, Hyderabad, Guwahati, Kolkata, pp.40-73.

Ibid, p. 40-73.

Ibid, p. 688.

Ibid, pp. 1-5.

Ibid, pp. 1-71.

Ibid, pp. 1-71.

Ibid, pp. 182-192.

Ibid, pp. 1955-1957, 1959-1960.

Ibid, pp. 285-307.

Ibid, pp. 44-45.

Ibid. Ibid. Ibid. Ibid.

IDSJ Working Paper 172, pp. 1-15.

International Journal of Plant Sciences, Vol. 6 Issue 2, pp. 274-276.

International Journal of Plant Sciences, Vol. 6 Issue 2, pp. 274-276.

International Journal of Plant Sciences, Vol. 6 Issue 2, pp. 274-276.

Israel, G. D. 1992. Determining Sample Size, PEOD-6, pp. 1-5.

J. 2015. Current trends of rubber plantation expansion may threaten biodiversity and livelihoods, Global Environment Change, 34, pp. 48-58.

Joseph, J., George K, T. and Dey, S.K. 2010. Report on the socio-economic impact of Natural Rubber cultivation under the block planting scheme in Tripura, MPRA Paper No. 52370, pp. 1-39.

Journal of Tribal Intellectual Collective India, Vol. 1 Issue 2 No 3, pp. 33-50. Debbarma,S.2009.RefugeeRehabilitationandLandAlienationinTripura,inW.

Kamaruddin, R. and Samsudin, S. 2014. The Sustainable Livelihoods Index: A Tool to Assess the Ability and Preparedness of the Rural Poor in Receiving Entrepreneurial Project, Journal of Social Economics Research, 1(6), pp. 108-117.

Kannan, M. 2013. Growth and Performance of Area, Production and Productivity of Natural Rubber in India, International Journal of Research in Commerce, Economic & Management, Volume no.3, Issue no.10, pp. 80-84.

Khatun, D. and Roy, B.C. 2012. Rural Livelihood Diversification in West Bengal: Determinants and Constraints, Agricultural Economics Research, Vol. 25 (No.1), pp. 115-124.

Khatun, D. and Roy, B.C. 2012. Rural Livelihood Diversification in West Bengal: Determinants and Constraints, Agricultural Economics Research, Vol. 25 (No.1), pp. 115-124.

Khatun, D. and Roy, B.C. 2012. Rural Livelihood Diversification in West Bengal: Determinants and Constraints, Agricultural Economics Research, Vol. 25 (No.1), pp. 115-124.

Krantz, L. 2001. The Sustainable Livelihood Approach to Poverty Reduction- An Introduction, Swedish International Development Cooperation Agency (Sida), Stockholm, Sweden, pp.1-27.

Krishnakumar, A.K. and Meenattoor.J, R. 2003. Restoration of denuded lands in the north eastern India through rubber plantations, LEISA INDIA, December, pp. 23-24.

Krishnakumar, A.K. and Meenattoor.J, R. 2003. Restoration of denuded lands in the north eastern India through rubber plantations, LEISA INDIA, December, pp. 23-24.

Krishnakumar, A.K. and Meenattoor.J, R. 2003. Restoration of denuded lands in the north eastern India through rubber plantations, LEISA INDIA, December, pp. 23-24.

Kurien, P.K. 2001. Sub-Marginal Rubber Cultivators: A study of livelihood issues of beneficiaries of 'Rubber to the Poor' project of MalanadDevelopment Society, Kanjirappally, Discussion Paper No. 32, Kerala Research Programme on Local Level Development, Centre for Development Studies, Thiruvananthapuram, pp.5-69.

Lagerqvist, Y. 2013. Imagining the Borderlands: Contending Stories of a Resource Frontier in Muang Sing. Singapore Journal of Tropical Geography, 34, pp. 57-69.

Liu, W., Hu, H., Ma, Y. and Li, H. 2006. Environmental and Socioeconomic Impacts of Increasing Rubber Plantations in Menglun Township, Southwest China, Mountain Research and Development, Vol. 26, No. 3, pp. 245-253.

Luangmay, D. and Kaneko, S. 2013. Expansion of Rubber Tree Plantation in Northern Laos: Economic and Environmental Consequences, Journal of International Development and Cooperation, Vol. 19, No. 3, pp.1-13.

M.D. Gupta, A.K. Gangopadhyay, T. Bhattacharya, M. Chakraborti (ed.) Forestry Development in North-East India, Omsons Publications, Guwahati and New Delhi, pp. 192-203.

Maithani, B.P. 2005. Shifting Cultivation in North-East India: Policy issues and options, Mittal Publication, New Delhi, pp. 30-50.

Maithani, B.P. 2005. Shifting Cultivation in North-East India: Policy issues and options, Mittal Publication, New Delhi, pp. 30-50.

Maithani, B.P. 2005. Shifting Cultivation in North-East India: Policy issues and options, Mittal Publication, New Delhi, pp. 30-50.

Maithani, B.P. 2005. Shifting Cultivation in North-East India: Policy issues and options, Mittal Publication, New Delhi, pp. 30-50.

Meti, S., Pradeep, B., Jacob, J., Meerabai, M. and Jessy, M.D. 2014. Spatio- Temporal Analysis of Rubber Area and its Association with Soil and Topography in Kanyakumari District, Rubber Science, 27(2), pp.182-192.

Meti, S., Pradeep, B., Jacob, J., Meerabai, M. and Jessy, M.D. 2014. Spatio- Temporal Analysis of Rubber Area and its Association with Soil and Topography in Kanyakumari District, Rubber Science, 27(2), pp.182-192.

Meti, S., Pradeep, B., Jacob, J., Shebin, S.M. and Jessy, M.D. 2016. Application of Remote Sensing and GIS for Estimating Area under Natural Rubber Cultivation in India, Rubber Science, 29(1), pp. 7-19.

Mohanakumar, S. 2013. Expansion of Natural Rubber Cultivation in Tripura: Impact on Landholding, Employment and Income, NLI Research Studies Series No. 107, pp. 1-46.

Mohanakumar, S. 2014. Political Economy of Natural Rubber Cultivation in Tripura,

Mors, W.B. and Rizzini, C.T. 1966. Useful Plants of Brazil, Holden-Day Inc., San Francisco, London, Amsterdam, pp. 1-7.

Mors, W.B. and Rizzini, C.T. 1966. Useful Plants of Brazil, Holden-Day Inc., San Francisco, London, Amsterdam, pp. 1-7.

Nath, D. and Bezbaruah, M.P. 2010. Income Augmentation in Small Farm Agriculture from Adoption of Rubber Production: A study in Northeast India, YSA, pp. 109-132.

Nath, T.K., Inoue, M. and Zoysa, M. 2013. Small-Scale Rubber Planting for Enhancement of People's Livelihoods: A Comparative Study in Three South Asian Countries, Society and Naturl Resources: An International Journal, 26:9, pp. 1066-1081.

Opeke, L.K. 1982. Tropical Tree Crops, John Wiley and Sons Ltd., Chichester, New York, Brisbane, Toronto, Singapore, pp. 216-240.

Opeke, L.K. 1982. Tropical Tree Crops, John Wiley and Sons Ltd., Chichester, New York, Brisbane, Toronto, Singapore, pp. 216-240.

Peet, R.K. 1974. The Measurement of Species Diversity, Annual Review of Ecology and Systematics, 5:1, pp.285-307.

Pensuk, A. and Shrestha, R.P. 2008. Effect of Paddy Area Conversion to Rubber Plantation on Rural Liveklihoods: A Case Study of Phatthalung Watershed, Southern Thailand, in W. Ongsakul (ed.) GMSARN International Journal, Vol.2, No.4, Greater Mekong Subregion Academic and Research Network, Thailand, pp. 185-190.

Perspective of Kerala, NRPPD Discussion Paper No. 38, Centre For Development Studies, Thiruvananthapuram, pp. 1-71.

Peter, K.V. 2002. Plantation Crops, National Book Trust, New Delhi, pp. 247-277. Gleason, C. 2005. The Biography of Rubber, Crabtree Publishing Company, pp. 1-32.

Polhamus, L.G. 1962. Rubber: Botany, Production, and Utilization, Leonard Hill [Book] Limited, London, Interscience Publishers, INC., New York, pp. 1-30, 347-392.

Polhamus, L.G. 1962. Rubber: Botany, Production, and Utilization, Leonard Hill [Book] Limited, London, Interscience Publishers, INC., New York, pp. 1-30, 347-392.

R.P. Athparia, and K. Jose (ed.) Tribal Development in Tripura, EBH Publisher, Guwahati, pp. 89-98.

R.P. Athparia, and K. Jose (ed.) Tribal Development in Tripura, EBH Publisher, Guwahati, pp. 89-98.

Raghavan, K.K. 2009. Rubber for tribal welfare in Tripura, in G.K. Bera, B.K. Medhi,

Raghavan, K.K. 2009. Rubber for tribal welfare in Tripura, in G.K. Bera, B.K. Medhi,

Rahman, S.A. 2006. The Beautiful India- Tripura, Reference Press Publication, New Delhi, pp. 191-200.

Rahman, S.A. 2006. The Beautiful India- Tripura, Reference Press Publication, New Delhi, pp. 191-200.

Rahman, S.A. 2006. The Beautiful India- Tripura, Reference Press Publication, New Delhi, pp. 191-200.

Raj, U., Hebbar, R., Ravishankar, H.M., Jacob, J., Ray, D., Meti, S., Shebin, S.M., & Pradeep, B. 2012. Geospatial Technology for Acreage Estimation of Natural Rubber and Identification of Potential Areas for its Cultivation in Tripura,

Project Report, Issue no. 1, National Remote Sensing Centre, ISRO, Govt. of India and Rubber Research Institute of India, Rubber Board, Min. of Commerce & Industry, Govt. of India, pp. 1-45.

Raj, U., Hebbar, R., Ravishankar, H.M., Jacob, J., Ray, D., Meti, S., Shebin, S.M., & Pradeep, B. 2012. Geospatial Technology for Acreage Estimation of Natural Rubber and Identification of Potential Areas for its Cultivation in Tripura, Project Report, Issue no. 1, National Remote Sensing Centre, ISRO, Govt. of India and Rubber Research Institute of India, Rubber Board, Min. of Commerce & Industry, Govt. of India, pp. 1-45.

Raj, U., Hebbar, R., Ravishankar, H.M., Jacob, J., Ray, D., Meti, S., Shebin, S.M. and Pradeep, B. 2012. Geospatial Technology for Acreage Estimation of Natural Rubber and Identification of Potential Areas for its Cultivation in Tripura, Project Report, Issue no. 1, National Remote Sensing Centre, ISRO, Govt. of India and Rubber Research Institute of India, Rubber Board, Min. of Commerce & Industry, Govt. of India, pp. 1-45.

Raj, U., Hebbar, R., Ravishankar, H.M., Jacob, J., Ray, D., Meti, S., Shebin, S.M. and Pradeep, B. 2012. Geospatial Technology for Acreage Estimation of Natural Rubber and Identification of Potential Areas for its Cultivation in Tripura, Project Report, Issue no. 1, National Remote Sensing Centre, ISRO, Govt. of India and Rubber Research Institute of India, Rubber Board, Min. of Commerce & Industry, Govt. of India, pp. 1-45.

Raj, U., Hebbar, R., Ravishankar, H.M., Jacob, J., Ray, D., Meti, S., Shebin, S.M. Pradeep, B. 2012. Geospatial Technology for Acreage Estimation of Natural Rubber and Identification of Potential Areas for its Cultivation in Tripura, Project Report, Issue no. 1, National Remote Sensing Centre, ISRO, Govt. of India and Rubber Research Institute of India, Rubber Board, Min. of Commerce & Industry, Govt. of India, pp. 1-45.

Raj, U., Hebbar, R., Ravishankar, H.M., Jacob, J., Ray, D., Meti, S., Shebin, S.M. and Pradeep, B. 2012. Geospatial Technology for Acreage Estimation of Natural Rubber and Identification of Potential Areas for its Cultivation in Tripura, Project Report, Issue no. 1, National Remote Sensing Centre, ISRO, Govt. of India and Rubber Research Institute of India, Rubber Board, Min. of Commerce & Industry, Govt. of India, pp. 1-45.

Rajasenan, D. 2010. Livelihood and employment of workers in rubber and spices plantations, NRPPD Discussion Paper- 6, pp. 1-68.

Rubber Board. 2016. Role of Rubber Board in Growth of Rubber Plantation in NE, The regional office, Agartala.

Rubber Board. 2016. Role of Rubber Board in Growth of Rubber Plantation in NE, The regional office, Agartala.

Rubber Board. 2016. Role of Rubber Board in Growth of Rubber Plantation in NE, The regional office, Agartala.

Rubber Board. 2016. Role of Rubber Board in Growth of Rubber Plantation in NE, The regional office, Agartala, Tripura.

Ruthenberg, H. 1980. Farming systems in the tropics, Oxford science publications-third edition, Oxford University Press, New York, Printed in Great Britain by The Thetford Press Limited, Thetford, Norfolk, pp. 301-305.

Saha, B. and Bahal, R. 2010. Livelihood Diversification Pursued by Farmers in West Bengal, Indian Res. J. Ext. Edu., 10(2), pp.1-9.

Saha, B. and Bahal, R. 2014. Livelihood Diversification Pattern among the Farmers of West Bengal, Economic Affairs, 59(3), pp.321-334.

Sarkar, M. 1986. Forestry Development in Tripura During the Five Year Plans, in

Sarkar, S. 2011. Rubber Plantation: A new hope for rural tribals in Tripura,

Sarkar, S. 2011. Rubber Plantation: A new hope for rural tribals in Tripura,

Sarkar, S. 2011. Rubber Plantation: A new hope for rural tribals in Tripura,

Sharma, G., George K., T. and Dey, S.K. 2014. Natural Rubber Sector in Tripura: Role of Institutional Innovations, Agricultural Economics Research Review, Vol. 27, pp 17-24.

Sharma, G., Joseph, J., George K, T. and Dey, S.K. 2011. Impact of Mahatma Gandhi National Rural Employment Guarantee Act on Rubber Block Plantation Scheme in Tripura, Agriculture Economic Research Review Vol. 24, pp. 525-530.

Simpson, E.H. 1949. Measurement of Diversity, Nature, Vol. 163, p. 688.

Sun, Z., Leinenkugel, P., Guo, H., Huang, C. and Kuenzer, C. 2017. Extracting distribution and expansion of rubber plantations from Landsat imagery using the C5.0 decision tree method, Journal of Applied Remote Sensing, 11(2), pp. 1-21.

TFDPCL. 2016a. Plan for Responsible Rubber wood and Bamboo Plantations Management (2013-14 to 2017-18), Agartala, Tripura, Prepared in consultation with Indian Institute of Forest Management, NEHRU NAGAR, BHOPAL, pp.1-201.

Tilahun and Teferie. 2015. Accuracy Assessment of Land Use Land Cover Classification using Google Earth, American Journal of Environment Protection, Vol. 4, No. 4, pp. 193-198.

Tripura,B.2013.TribalQuestioninTripura:DialogueBetweenItsPastandPresent,

TRPCL. 2016. Document on Tripura Rehabilitation Plantation Corporation Limited, Agartala, Tripura.

TRPCL. 2016. Document on Tripura Rehabilitation Plantation Corporation Limited, Agartala, Tripura.

USGS. EarthExplorer. Available online: http://earthexplorer.usgs.gov/.

Verheye, W. 2010. Growth and Production of Rubber, in Verheye, W. (ed.) Land Use, Land Cover and Soil Sciences, Encyclopedia of Life Support Systems (EOLSS), UNESCO-EOLSS Publishers, Oxford, UK, pp.1-20.

Vinod, K.K., Meenottoor, J.P., Priyadarshan, P.M., Pothen, J., Chaudhuri, D., Krishnakumar, A.K., Sethuraj, M.R., Potty, S.N. 1996. Early performance of some clones of Hevea Brasiliensis in Tripura, Indian Journal of Natural Rubber Research, 9(2), pp. 123-129.

Viswanathan, P.K. 2008. Emerging Smallholder Rubber Farming System in India and Thailand: A Comparative Economic Analysis, Asian Journal of Agriculture and Development, Vol. 5, No. 2, pp. 1-16.

Viswanathan, P.K. and Bhowmik, I. 2014. Compatibility of Institutional Architecture for Rubber Plantation Development in North East India from the Comparative Perspective of Kerala, NRPPD Discussion Paper No. 38, Centre For Development Studies, Thiruvananthapuram, pp. 1-71.

Viswanathan, P.K. and Bhowmik, I. 2014. Compatibility of Institutional Architecture for Rubber Plantation Development in North East India from the Comparative

Viswanathan, P.K. and Bhowmik, I. 2014. Compatibility of Institutional Architecture for Rubber Plantation Development in North East India from the Comparative Perspective of Kerala, NRPPD Discussion Paper No. 38, Centre For Development Studies, Thiruvananthapuram, pp. 1-71.

Viswanathan, P.K. and Bhowmik, I. 2014. Compatibility of Institutional Architecture for Rubber Plantation Development in North East India from the Comparative Perspective of Kerala, NRPPD Discussion Paper No. 38, Centre For Development Studies, Thiruvananthapuram, pp. 1-71.

Viswanathan, P.K. and Bhowmik, I. 2014. Compatibility of Institutional Architecture for Rubber Plantation Development in North East India from the Comparative Perspective of Kerala, NRPPD Discussion Paper No. 38, Centre For Development Studies, Thiruvananthapuram, pp. 1-71.

Viswanathan, P.K. and Bhowmik, I. 2014. Compatibility of Institutional Architecture for Rubber Plantation Development in North East India from the Comparative Perspective of Kerala, NRPPD Discussion Paper No. 38, Centre For Development Studies, Thiruvananthapuram, pp. 1-71.

Viswanathan, P.K. and Shivakoti, G.P. 2008. Adoption of rubber-integrated farm livelihood systems: contrasting empirical evidence from the Indian context. J For Res, 13, pp. 1-14.

Vongkhamheng, C., Zhou, J.H., Beckline, M. and Phimmachanh, S. 2016. Socioeconomic and Ecological Impact Analysis of Rubber Cultivation in Southeast Asia. Open Access Library Journal, 3: e2339.

Vongkhamor, S., Phimmasen, K., Silapeth, B., Xayxomphou, B. and Petterson, E. 2007. Key Issues in Smallholder Rubber Planting in Oudomxay and Luang Prabang Provinces, National Agriculture and Forestry Research Institute Vientiane, Lao PDR.

Welfare for Scheduled Tribes and Scheduled Castes, Government of Tripura, Agartala, Print Best, Agartala, Tripura, pp. 9-39.

2

Geographical Background of Tripura

2.1 Introduction

Tripura a continuation of the Himalayan ranges is covered by hilly terrain locally termed as 'tilla'. intercepted by fertile valley locally termed as "Lunga." The state has experienced huge influx of immigrants since its inception which has altered the demographic composition of the state. Tripura is predominantly agrarian as more than 75 % of the state's total workforce is dependent on agriculture for their sustenance With hilly terrain, fertile lowlands , abundant rainfall , rivers and streams the state is rich in biodiversity. Socio- economically though relatively backward , Tripura has a rich history being suled by the native kings with a dynamic culture, who adopted Bengali as their official language in the 19th century.

2.2 Brief History

The history of Tripura is shrouded in ancient Indian accounts of myths and legends. The name of the state of Tripura is believed to be name in honor of king Tripur who was the fortieth descendant in the Manikya dynasty. As per *Rajmala* (Chronicle of the kings of Tripura) the name of the state was given from the words *Tui* (water) and *Pra* (near) which literary mean land adjoining water. Other scholars believe that the name of the state is the sanskritized from of *Tipra* tribe, the tribe who ruled the state from which the ruling family belongs. Scholar like W.W. Hunter opines that the name Tripura is in honor of the Tripureswari temple located at Udaipur built around AD 1501 by Dhanya Manikya who ruled the state from AD 1490 to 1515 (Bera, 2010).

The history of Tripura is described in the *Rajmala* and historical records recorded from 1407 A.D by both the *Rajmala* and the Muhammedan historians. There are references of Tripura dating back to the Mahabharata, the Puranas and the pillar inscription of emperor Ashoka. Historical account confirms that Tripura was an ancient kingdom. Tripura has been rule by many Maharajas continuously for about thirteen hundred years (Chatterjee, 1984).

Evidence from archaeological survey proves that a large portion, particularly south-western part of the present state of Tripura was politically attached to the kingdoms of Eastern Bengal known as Samatata Vanga and Harikela. However, from the historical records of the Gupta dynasty, Naths dynasty, Ratas dynasty, Khadgas dynasty, Chandras dynasty, Varmans dynasty, the kingdoms of Burma and Pattikara, the Devas kingdom of Nadiya and the Devas kindom of Srihatta suggest that the present state of Tripura was not found in the ancient period. Most of the western and southern parts of present Tripura were politically attached to Eastern Bengal, however, in ancient period, some tribes (mainly the Tipras who were numerous and well organized) inhabited the hilly north-eastern parts of present Tripura but without any political importance. The Tipra tribes were able to establish a kingdom in this region (the hilly region of Tripura) after the fall of the Devas dynasty in 1257 A.D, which later came to be known as the Tripura kingdom (Chaudhuri, 1980).

As per *Rajmala* the kingdom's boundary extended up to the present Nowgong district of Assam, the capital of this kingdom was in present Cachar district (particularly in north Cachar hills) on the bank of the Barak River. Later it shifted towards south-west to present Dharmanagar on the bank of the Juri river; from Dharmanagar the capital shifted and moved south-westward and finally got located in the present capital of Tripura i.e. Agartala (Chaudhuri, 1980). The kingdom has faced frequent attack by the Mughal force but the Rajas repulsed these attacks bravely for about three centuries. During the British reign Tripura remained a native kingdom ruled by Maharajas of Manikya dynasty. However, after the death of Maharaja Bir Bikram Kishore Manikya on the 17th of May 1947, Her Highness Maharani Kanchan Prova Devi formed a Regency Council in order to continue with the administrative function of the kingdom. With Indian independence the political scenario change and on 15th of October 1949, the kingdom of Tripura was merged with the Union of India (Chaudhuri, 1980; Chatterjee, 1984).

After joining with the Union of India, Tripura became a Part 'C' state administered by the Chief Commissioner and later became a Union Territory in 1956.

Tripura attained statehood on the 21st of January 1972 under the North Eastern Regions Re-organisation Act, 1971 (Chatterjee, 1984). However, the partition of British India in 1947 and integration of Tripura with the Indian union has forever changed the history of the erstwhile princely kingdom due to large flow of population from adjoining East Pakistan (now Bangladesh).

Before attaining statehood, on 1 September 1970, the state of Tripura was divided into three districts i.e. North Tripura, West Tripura and South Tripura districts (Chatterjee, 1984). After attaining statehood in 1972, Dhalai district was created in 1995 which was previously a part of the North Tripura district (http://dhalai.gov.in/new/.html). The state had four districts till 2012. However during 2012 the Government of Tripura created another four new districts from the old three districts i.e. Unokoti district from previous North Tripura district, Khowai and Sipahijala districts from previous West Tripura district and Gomati district from previous South Tripura district for the purpose of effective administration. 4 present the state has eight districts in total i.e. North Tripura, Unokoti, Dhalai, Khowai, West Tripura, Sipahijala, Gomati, and South Tripura districts and one Tripura Tribal Areas Autonomous District Council (TTAADC) created under the sixth schedule of the constitution (DESPD, 2012). The former four districts map of the state is shown in figure 2.1 and the present eight districts map of the state is shown in figure 2.2.

2.3 Geographical setting

2.3.1 Location and Area

Tripura is situated in the south-western part of the Northeast India and holds a strategic location as it is bounded on three sides i.e. in the north, south, and west by neighboring country Bangladesh. The state is connected to rest of India through the north-eastern part of Assam and Mizoram states. It is located between 22° 56′ to 24° 32′ N latitude and 91° 09′ to 92° 20′ E longitude. Consequently, partition of the country has severed Tripura of its direct link with eastern India viz. Kolkata making it a remote land-locked state a strategic location bounded on three sides directions by international boundary (GoT, 2007).

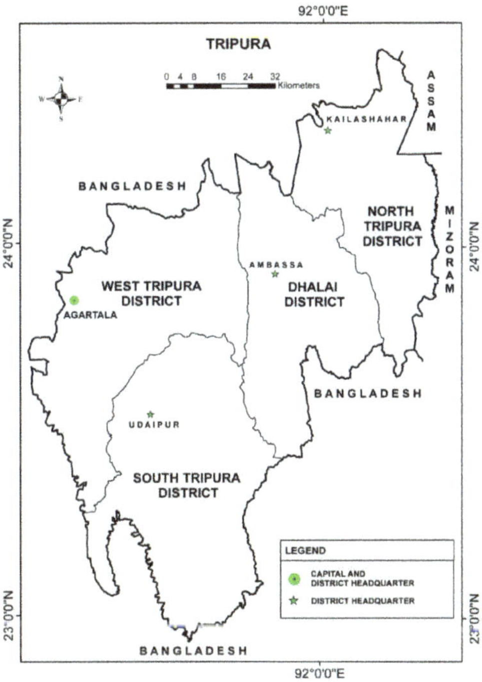

Fig. 2.1: Four districts of Tripura before reorganization.

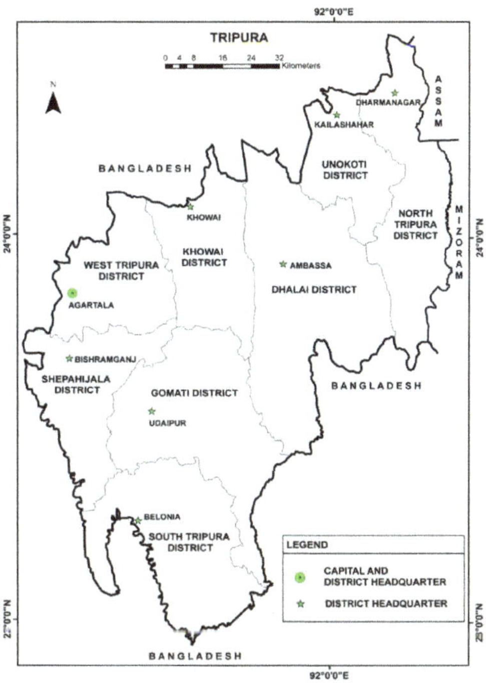

Fig. 2.2: Eight districta of Tripura after reorganization.

The total geographical area of the state is 10,486 square kilometres (Census of India, 2011) which constitute about 0.32 percent of the total geographical area of the country (India) and is the third smallest state of the country and second smallest state in Northeast India next only to Sikkim. The state has a boundary of about 1,018 kilometres of which 856 kilometres (84 percent) is an international boundary (i.e. with Bangladesh) and only 162 kilometres (16 percent) of its boundary is a national boundary sharing about 109 kilometres with Mizoram and 53 kilometres with Assam respectively. Of the total geographical area, about 60 percent of the geographical area is under forested hill ranges (GoT, 2007).

2.3.2 Physiography

Physiographically, the area of the state represents the western fringe of the typical ridge and valley mountain belt of the Indo-Burma Ranges (Bhattacherjee and Singh, 1995). Geomorphologically, the state is characterized with rugged topography. The terrain consists of six parallel hill ranges viz. Deotamura, Baramura, Atharamura, Longtarai, Skkanklong and Jampui ranges which run in a north to south direction and between each hill ranges lies river valleys expressing topography that exhibit anticlinal structures and synclinal valleys. The valleys are not very broad and are characterized with undulating or flat features extremely fertile ideal for agriculture (Chatterjee, 1984).

The state can be further divided into three broad topographic categories viz.

1. hill ranges consisting of the six major ranges, (2) undulating highland with narrow and broken plateau structure and (3) lowlands and river flats (Laskar, 1966, cited in Laskar, Dadhwal and Prasad, 1983). The hill ranges and uplands are locally termed as *muras* and *tillas* respectively while the valleys and lowlands are termed as *lungas*. The ground rises from west to east; however elevation of each hill ranges are gradually higher in the south and lower in the north with the highest peak situated on the Jampui ranges i.e. 937 metres (m). The altitude of the state varies between 10 m to 600 m above mean sea level (m.s.l) (Kesari, 2011).

The hill ranges and hillocks of the state was once covered by dense forests with timber forest, bamboo bushes etc which constitute rich natural wealth, but the forest area has been shrinking continuously due to encroachments by the increasing population, practice of *jhum* cultivation and procurement of wood for commercial purposes (Saigal, 1978; Chatterjee, 1984). *Tilla* lands are considered suitable for *jhum* cultivation, forest plantation, orchard, tea cultivation and rubber plantation etc. *Lunga* and plain land on the other hand are fertile and rich in humus deposited by rivers and streams from the hill ranges. The *lunga* and plain lands are suitable for plough cultivation generally known as settled agriculture.

Geology

Geologically, the hill tracts consist of a sub-stratum of tertiary rocks, which has been covered by later alluvial deposits. The sub-stratum has been exposed during the Himalayan uplift in Eocene times. There are three series of sedimentary rocks viz. Barail series (consists of fairly hard fine grained and sand stones), Surma series (consists of massive sandstone and medium to coarse grained) and Tipam series (consists of soft coarse grained sand stones). The three series of sedimentary rocks lays one above the other. The arrangement of rock stratum indicates instability as the state falls under the earthquake prone zone (Chatterjee, 1984).

2.3.3 Climate

Climatically, the state experience sub-tropical, humid and hot climatic condition. The principal seasons of the state are pre-monsoon season between March to May, monsoon or rainy season between June to September which bring cyclones, hailstorms, thunderstorms and lightening in its wake causing damages to houses, crops, trees and sometimes deaths of animals and human, autumn or post monsoon season between October to early November and winter and spring season between December to February where winter continues up to January and spring in the month of February (Gan-Chaudhuri, 1980; Laskar, Dadhwal and Prasad, 1983; Chatterjee, 1984).

The average maximum temperature during summer (May and June) is 35 degree Celsius and the average minimum temperature during winter (December and January) is 10 degree Celsius. Sometimes the temperature falls to 5 degree Celsius during winter (Kesari, 2011). There is however difference of temperature between the hills and plains. The topographic features play an important role in influencing the climatic condition of the state, where the plains are hotter and humid in comparison to the hills, which have a salubrious climate.

The state receives most of its rainfall from the south-west monsoon, from the month of June which last till about the last part of September. Rainfall is not uniform as the north and east parts of the state receive maximum rainfall while rainfall decreases towards the south and western parts of the state. The hill ranges having north-south alignment affects both the temporal and spatial distribution of rainfall as during south-west monsoon one part of the region or a river basin experience deluge and consequent floods while other part of the state may experience drought (Chatterjee, 1984). The average annual rainfall in the state is 1,600 to 2,500 mm with maximum rainfall occurring during the south-west monsoon period (Kesari, 2011).

Table 2.1: Annual rainfall in Tripura (2011-2015)

Year	Annual Rainfall (mm)
2011	1,861.6
2012	1,881.3
2013	1,952.5
2014	1,870.1
2015	2,229.7

Source: Agriculture Department of Tripura, cited in DESPD, 2016.

2.3.4 Rivers

In Tripura there are nine major rivers namely Juri, Manu, Deo, Dhalai, Khowai, Gumati, Haora, Muhuri and Burigang rivers. These rivers carry enormous quantities of water especially during the south-west monsoon season bringing happiness among the farmers but sometimes cause heavy floods. As these rivers are rain fed rivers the water is not adequate for agricultural purposes throughout the years (Chatterjee, 1984).

River Juri originates in Jampui range, Manu river in Kohoisib peak in Sakkanklong range and Deo and Dhalai are the tributaries of Manu river one on its right and left bank respectively. Khowai rivers emerges from Longtarai and Atharamura hill ranges, Gumati river is formed by joining of two rivers namely Sarama and Raima with the former originating in the Atharamura range and the latter in the Longtarai range. Gumati river is considered as the main principal river of the state as it runs almost through the middle part of the state. Both Haora and Burigang rivers emerge from Baramura hill range (Chatterjee, 1984).

Each valley contains a stream in which smaller order streams join carrying water from the hills. Juri, Khowai, Manu, Deo and Dhalai rivers flow from south to north direction. Gumati and Muhuri rivers flow southwards while Haora and Burigang rivers flow from east to west. All the major rivers later join rivers of Bangladesh. Topography influences the drainage pattern where longitudinal streams occupy the northern half of Tripura while dendratic pattern is found in the west and southern part of the state (Chatterjee, 1984).

2.3.5 Soil

Geology, climate, and geomorphological processes play a significant role in shaping the composition and structure of the soil. The soil of the state is sandy loam to loamy sand, clay loam to pure clay or lateritic.

According to the physiography features, the soils of the state are categories into (1) Soil of the uplands and (2) Soil of the lowlands.

Soil of the uplands mainly consists of disintegrated bed rocks i.e. sandstone and shales. Soil developed from sandstones are coarse in texture, reddish brown to brown in colour, poor in humus and lime content. On the other hand, soils developed from shales are medium to fine-grained in texture, dark brown to dark grey brown in colour, acidic and poor in permeability and lime content. Uplands soils do not hold water as the soil is characterized with loose and porous materials. Therefore during the rainy season, mineral particles of the soil are drained making the soil mineral deficient. Although the uplands are covered with forests yet it does not offer good soil for growing of crops and therefore *jhum* cultivation is practice in the uplands (Chatterjee, 1984) which had initially emerged as a response to the prevailing geo- environmental and cultural factors but at present needs to be checked due to environmental problems, the growth of population, and modern lifestyle etc are leading to smaller *jhum* cycles resulting to increasing deforestation and associated problems.

Soil in the lowlands is mostly alluvial in nature. It is well drained and deposited as sedimentary alluvium brought down by rivers originating from the surrounding hills ranges. The soil is deep, medium to fine grained and light grey in colour. It covers river valleys and flood plains areas locally termed as *lungas*. Soils of the lowlands are intensively cultivated due to the presence of organic matter in the soil (Chatterjee, 1984). The hill slopes are well drained and fertile, in recent years rubber plantation have come up in the hill slopes and foothills throughout the state as the agro-climatic and soil conditions for rubber cultivation are found ideal here.

2.3.6 Vegetation

Physiography, climate, rainfall and soil condition plays an important role in determining the vegetation of any area. In Tripura, forest with thick vegetation cover are found in most of the elevated flat land, hillocks and the high hills as rainfall are well distributed, and the region receives sufficient sunlight which favors luxuriant growth of vegetation. Varieties of bamboo species as well as other plants are found here. Cane and grasslands mainly cover the plains and the small hillocks and lowlands. The natural vegetation are often considered as an important source of livelihood for the people living in the forested tracks as they use the forest products as food, fuel, tools, weapon, basketry, house construction medicinal purpose etc. Important species of trees here are Kanak, Karai, Jam, Gamir, Sonal, etc. which are found all over the state (Laskar, Dadhwal and Prasad, 1983; Chatterjee, 1984). Forests occupy most parts of the hills and are almost absent in the valley areas. Sal (*Shorea robusta*) forest, Garjan (*Dipterocarpus turbinatus*) forest, Bamboo (*Bambusa indica*) forest and miscellaneous forest are the main types of forest. All these types are found either separately or mixed together and grow all over the state. Sal forests are mainly distributed in a narrow belt in Udaipur, Sonamura and Belonia sub-divisions. There are two sub-types of Sal forest, one known as highlands type as it grow in highlands and another is known as paddy field type as it grows near paddy fields. Garjan forests exist in small patches in Belonia, Sabroom, Udaipur and Sadar sub-divisions. Bamboo forests occupy vast areas in the hilly tracts and also forms impenetrable jungle in association with other plants. Miscellaneous forests are mostly dry or moist deciduous types and found all over the state (Saigal, 1978; Chatterjee, 1984).

Since few decades, some new species of trees which are not native to this land have been introduced of which rubber, teak, cashew-nut and eucalyptus are the important ones. Teak plantation has been introduced mainly for the enrichment of forest species. Cashewnut plantation has been implemented for soil conservation purposes. Eucalyptus has been planted to increase forest trees as it is a fast growing species. Rubber plantation has been introduced for soil conservation measures and also for commercial purposes (Chatterjee, 1984).

As mentioned the state of Tripura has rich forest resources, but due to the population pressure, *jhum* cultivation, large-scale un-systematic and unrestricted felling of valuable forest trees have not only destroyed and reduced the forest resources but also accelerated soil erosion (Saigal, 1978; Chatterjee, 1984).

Of the total geographical area, forests occupy about 6,294 sq km. However as per report of the Forest Survey of India during 2015, the total forest cover in the state is 7,866 sq km covering 74.97 percent of the geographical area of the state with 109 sq km as very dense forest, 4,641 sq km as moderately dense forest and 3,116 sq km as open forest (Forest department of Tripura, 2015, cited in DESPD, 2016).

2.3.7 Mineral resources

Tripura is not so rich in term of mineral resources. Some of the mineral deposit include clay, limestone, lignite coal and ochre however the quality and reserve of these minerals are poor for commercial utilization. Of these, clay is considered the most important deposit which is of workable value followed by limestone and glass sands. Recently natural gas deposits have been discovered and are considered important natural resource. Natural gas bearing structure is identified in the

Baramura-Atharamura hill ranges and in South District. Systematic exploration and scientific exploitation of minerals are needed (Saigal, 1978; Chatterjee, 1984).

2.4 Population

The partition of India in 1947 has changed the whole scenario of the state, which has led to both demographic and socio-economic changes (Bareh, 2001; Debbarma, 2009; Bhaumik, 2015). As a result of large influx of refugees into the small state from erstwhile East Pakistan (now Bangladesh) during partition has created pressure on land resource (DESPD, 2016). Before the merger of the princely state of Tripura with the India union, the indigenous heterogeneous groups of *Borok* tribal groups mainly inhabited the state and they were in majority but at present have been reduced to minority (Bhattacharjee, 1989) and experience alienation from land, marginalization in jobs, professions, and politics, this was further accentuated with the next spike of large-scale immigration during 1967 and during the Bangladesh Liberation war (1971) (Bareh, 2001). The 1971 census reflects that Tripura decadal growth rate (1961-71) was higher by 11.68 percent than the average growth rate of India for the same decade (i.e. 36.28 percent and 24.60 percent respectively) suggesting not only natural growth of population but also migration. In fact, about 1 million refugees from present Bangladesh migrated to this adjoining area during the Bangladesh Liberation war. As a result of large influx of immigration, the indigenous communities are not only reduced to a minority but as already mentioned suffering from land alienation. This demographic growth and changes has its impact on the economy of the state (DESPD, 2016).

Debbarma (2009) states that there were Bengalis in Tripura before the partition of British India but their numbers were insignificant. Partition and the later integration of Tripura into Indian union in 1949 saw large influx of population who started migrating from nearby East Pakistan. Bengali speaking populations are at present the dominant group of the state and play a vital role in the economy and political affairs of the state. The indigenous population has become a minority in its own land accompanied with loss of livelihood opportunities.

This demographic change in composition has gradually emerged as a vexed issue in Tripura. Tripura is also the only state in India where successive migration has reduced its original residents to a minority in their own land (Bareh, 2001; S.R Bhattacharya, 1989; S.B Bhattacharya, 1990 as cited in Tripura, 2013). .

As depicted in table 2.2, the share of ST population has been declining while the population share of others (mainly Bengali speaking population) has been increasing. As per census 2011, ST population constitutes about 31.76 percent of the state's total population.

At present the state of Tripura is densely populated. As per census of India 2011, Tripura is home to 36, 73, 917 persons, out of which 18, 74, 376 are males and 17, 99, 541 are females; having a density of about 350 persons per square kilometres. Of the total state populations, 11, 66, 813 (31.8 percent) consists of ST population and the rest 25, 07, 104 persons are non-tribal population. About 73.8 percent of the total population inhabits the rural area mostly consisting of ST population (Census of India, 2011).

Table 2.2: Share of ST and general population to the total population of Tripura (1901-2011)

Census year	Total population	ST population	ST population (in %)	General population	General population (in %)
1901	1,73,325	91,679	52.89	81,646	47.11
1911	2,29,613	1,11,308	48.48	1,18,305	51.52
1921	3,04,437	1,71,610	56.37	1,32,827	43.63
1931	3,82,450	1,92,240	50.27	1,90,210	49.73
1941	5,13,010	2,56,991	50.09	2,56,019	49.91
1951	6,45,707	2,37,953	36.85	4,07,754	63.15
1961	11,42,005	3,60,070	31.53	7,81,935	68.47
1971	15,56,342	4,50,544	28.95	11,05,798	71.05
1981	20,53,058	5,83,960	28.44	14,69,098	71.56
1991	27,57,205	8,53,345	30.95	19,03,860	69.05
2001	31,99,203	9,93,426	31.05	22,05,777	68.95
2011	36,73,917	11,66,813	31.76	25,07,104	68.24

Source: Census reports, cited in DESPD, 2016

Tribal composition-

In Tripura, there are 19 ST groups enlisted by the Government of India i.e. Tipra/Tripuri/Tripura, Riang, Jamatia, Noatia, Uchai, Lushai, Mog, Kuki, Halam, Chakma, Garo, Khasi, Bhutia, Lepcha, Bhil, Munda, Orang, Santal and Chaimal (Debbarma, 2009). Among these Tripuri, Reang, Jamatia and Chakma are considered the major ST groups of the state. The indigenous tribes of the state belong to Indo- mongoloid racial stock (Bhat and Bhargava, 2006) and speak Tibeto-burman languages that may be broadly categorized into two major linguistic groups i.e. the *Kokborok* speaking group consisting of the Tripuri, Reang, Jamatia, Noatia sub-tribes and the Kukichin speaking group consisting of the Uchoi, kuki, Halam, Lushai, Mog sub-tribes.

Table 2.3 indicated that the Tipra/Tripuri/Tripura community has the largest share among ST population and account for about 50.76 percent followed by the Reang tribe (16.13 %), Jamatia tribe with 7.14 percent, Chakma tribe with 6.84 percent, and Halam tribe with 4.90 percent. The Noatia, Mog, Garo, Munda, Orang and the Generic (not specific), each account for above 1 percent to the total ST population. The tribes which have less than 1 percent share are Kuki (0.94 %), Lushai (0.46 %), Bhil (0.27 %), Santhal (0.25 %), Uchai (0.21 %), whereas, Chamal, Khasia, Lepcha and Bhutia have negligible share in population to the total ST population. Some of these tribes are occupying the forested tracks of the state and are still in the process of social evolution, practicing primitive agricultural practices like *jhum* cultivation in the hill slopes (Gupta, 2007). It is for these ST populations that the state government through its different agencies are encouraging rubber plantation so that they can be brought within the ambit of settled agriculture.

Table 2.3: Types of ST population and its share to total ST population in Tripura (2011 Census)

Name of the ST group	Total population	ST population (in %)	Rank
Tipra/Tripuri/Tripura	5,92,255	50.76	1
Reang	1,88,220	16.13	2
Jamatia	83,347	7.14	3
Noatia	14,298	1.23	9
Uchai	2,447	0.21	16
Kuki	10,965	0.94	12
Halam	57,210	4.90	5
Lushai	5,384	0.46	13
Bhutia	28	0.002	20
Lepcha	157	0.01	19
Khashia	366	0.03	18
Chakma	79,813	6.84	4
Mog	37,893	3.25	7
Garo	12,952	1.11	10
Munda/Kaur	14,544	1.25	8
Santhal	2,913	0.25	15
Orang	12,011	1.03	11
Bhil	3,105	0.27	14
Chamal	549	0.05	17
Generic	48,356	4.14	6
Total	11,66,813	100	

Source: Census of India, 2011.

As depicted in table 2.4, of the total population, about 47 percent of the total population of the state is noticed inhabiting West Tripura district which also has the largest share of population. West Tripura district is followed by South Tripura district (23.74 %), North Tripura district (18.89 %) and Dhalai district (10.30 %).

Regarding ST population, it is Dhalai district that has the highest share with 55.68 percent of the total reported as ST population, followed by South Tripura district which accounts for about 39 percent of the total ST population. West Tripura and North Tripura districts have the least share of ST population with about 25.18 percent and 25.86 percent respectively.

Table 2.4: District wise distribution of ST population in Tripura (2011 Census)

Districts	Total population	Share in %	ST population	Share in %
West Tripura	17,29,451	47.07	4,35,534	25.18
South Tripura	8,72,289	23.74	3,41,245	39.12
North Tripura	6,93,947	18.89	1,79,426	25.86
Dhalai	3,78,230	10.30	2,10,608	55.68
Total	36,73,917	100	11,66,813	31.76

Source: Census of India, 2011.

2.5 Economy

Tripura in considered as an agricultural state and thus economy of the state is basically agrarian in nature. For most of the people of the state, agriculture is considered the principal occupation although the areas available for agriculture purposes are restricted to small areas. Agriculture and allied activities are considered as the backbone of the state's economy (DESPD, 2016).

Being an agricultural state, the state per capita income has always been low. Industrialization is lacking due to geographical isolation, poor road and railway connectivity. On the other hand, the un-organized manufacturing and service activities are more dominant in comparison to the organized sector in the state. These constraints are also considered as the main reasons for the slow economic growth of the state (DESPD, 2016).

Two types of agriculture are practiced in the state viz. (1) *jhum* cultivation, mostly practiced by the ST population inhabiting the hills and hill slopes and (2) settled agriculture, mostly practiced by the people inhabiting the valleys and plain areas (DESPD, 2016). The practiced of *jhum* causes serious threat to the forests and recently the state government has introduced some new species of trees (including rubber) as a conservation measures to check further degradation of forest areas.

As already mentioned new species of trees including cultivation of rubber trees was introduced for conservation measures. Rubber plantation has been a success (table 2.5). Among the principal crops in the state, rubber plantation is ranked second in coverage of area and production after rice cultivation. This suggests that cultivation of rubber trees is gaining popularity among the masses and this is mainly due to economic opportunities which rubber plants offer in comparison to others crop in the state.

Table 2.5: Area and production of principle crops in Tripura (2015-16)

Sl. No.	Item	Area (in ha)	Production (in Mt)
1	Rice (Aush, Aman, Jhum, Kharif, Boro)	2,75,193	7,99,437
2	Rubber plantation	74,335	49,943
3	Wheat	182	364
4	Foxtail/Kaon	100	80
5	Cotton (Bale=170 Kg)	904	1383 (in Nos.)

Sl. No.	Item	Area (in ha)	Production (in Mt)
6	Jute (Bale=180 Kg)	650	5,675 (in Nos.)
7	Mesta (Bale=180 Kg)	627	5,217 (in Nos.)
8	Total pulses (kharif & rabi)	22,327	13,925
9	Total oilseeds (kharif & rabi)	11,727	9,651
10	Maize (kharif & rabi)	8,646	11,578

Source: DESPD, 2016.

The contribution of agriculture and other allied activities in the gross domestic product (GDP) of the state is low but its contribution has been increasing slowly in the recent decades. During 2014-15 agriculture and other allied activities sectors has contributed about 26 percent to the state GDP. The share of agriculture in the state's GDP is about 33 percent in 2014-15.

The GDP (with 2011-12 base year) have increased from Rs. 19,208.41 crores in 2011-12 to Rs. 21,663.20 crores in 2012-13 which has further increased to Rs. 25,592.83 crores in 2013-14 and to Rs. 29,666.62 crore in 2014-15.

With liberalization of the Indian economy in 1990s and associated reforms the Indian economy started improving accelerating industrial growth and maintaining the inclusive growth as experienced by the country, but this has not given much boost for industrialization in the state. During the 12th plan period (2012-17), although the economy of the state was impressive but the result of slowdown of the national economy has generated further fiscal stretch (DESPD, 2016).

2.6 Socio-economic scenario

During the Maharajas eras, no doubt the development pace was slow but the economic condition was much better though stagnant and semi-feudal. As the population was small and traditional with low literacy rate and lower aspiration level the state did not confront with any serious resource crunch, no pressure on land, no serious unemployment challenge that could create serious threat to the social structure and economy of the state. The princely state although backward compared to other parts of India was self-sufficient in food. However, due to population growth and lack of accessibility created with partition of the country, Tripura became landlocked isolated state difficult to reach from mainland India and could not achieve any significant diversification in creating income and employment opportunities. There is also marked difference in the level of economic and development between the hills and the plain regions of the state (Bareh, 2001).

The partition of India in 1947, has not only affected demographic changes but has it impact on resource allocation, chocked off transport and communication network and the political setup of the state. Extremism and insurgency issues have emerged which to some extend has hindered the economic development process especially in the last two decades (DESPD, 2016).

As per current survey on the land under cultivation conducted by agriculture department of the state, 24.33 percent of the total geographical area is under agriculture. Although about 60 percent of the population totally dependent on agriculture and about 50.83 percent of the total population are directly dependent on agricultural activities for their livelihood and are engaged mostly in intensive subsistence type of farming, The share of small and marginal farmers to the total farmers in Tripura is as high as 96 percent against 78 percent reported by the country. Thus agricultural sector is considered as the main backbone of the state's economy (DESPD, 2016).

As already mentioned the state is characterized by high incidence of poverty, low per-capita income, low capital formation, inadequate infrastructure facilities, geographical isolation, and lack of communication, inadequate exploitation and use of forest and mineral resources, low progress in industrial field and high unemployment problem. These have disrupted the process of economic development of the state. Although the socio-economic conditions and available lands are favorable for development of various industries in the state like promotion of natural gas, rubber based industries, handloom and handicraft industries, tea industry through modern technology, eco-tourism, trade between neighboring countries etc which will not only improve the economic prosperity of the people here as it will generate employment opportunities (DESPD, 2016).

In Tripura, the population is characterized by social diversity where about 74 percent of the population inhabits the rural areas. Climatic condition with high rainfall and fertile soil offers the state with great potential for plantations. The state has introduced a variety of plantation crops, the important amongst them are tea, rubber, cashew nut, pineapple and oranges while jackfruit, coconut, banana, lemon and areca nut. Besides this, varieties of vegetables, spices are also grown for commercial purpose, the state also has a great potential for high value of horticultural crops, floriculture, medicinal, aromatic plants etc and food processing. Among the plantation crops, 'rubber and tea plantation activities have a special socio-economic significance in rehabilitating the ST *jhumias* population of the state (DESPD, 2016).

The introduction of rubber plantation however has changed the dynamics of a relatively classless society of the ST population. The ST rubber growers, initially a mix of poor *jhumias* and small-time ST government employees, teachers and petty traders have now emerged as a powerful social group in Tripura's tribal society, but they greatly depend on the state and the Bengali industrial class for processing latex into sheets and promotion of rubber products through transport subsidies, sale and marketing to industries of West Bengal where they get the best price (Bhaumik, 2015).

In the North-Eastern region, Tripura has been ranked the highest in chronic unemployment in rural sector as thousands of ST remain landless, despite the government's scheme for restoration of alienated land to them (Bareh, 2001). Hence, the ST populations here are often in a disadvantageous position as they are still in the process of social evolution and often find it difficult to be associated with a more modern monetized economy. However, both the central government and the state government have introduced various scheme for resolving the land alienation problems and socio-economic development of especially the weaker section of the ST population of the state.

Consequently rubber plantation have been introduced in the state with the objective of not only a conservation measure for soil degradation and afforestation but also for improving the socio-economic development of the *jhumias* and also minimizing the land alienation problems and bringing the *jhumias* within settled agriculture by slowly transforming the social and economically marginalized section of population. Consequently, some of the marginalized section practicing *jhum* cultivation have transformed to rubber plantation owners. Nevertheless, before we examine the impact of rubber plantation, it becomes important to examine the growth and distribution of area under rubber plantation in the state which is highlighted in the next chapter.

References

Bera, G.K. 2009. Introduction, in G.K. Bera, B.K. Medhi, R.P. Athparia, and K. Jose (ed.) *Tribal Development in Tripura,* EBH Publisher, Guwahati, pp. 1-16.

Chatterjee, S.N. 1984. *Tripura: A Profile,* Inter- India Publications, New Delhi, pp. 1- 61.

Chaudhuri, N.R.R. 1980. The Historical Past, in J. Gan-Chaudhuri (ed) *Tripura: The Land and its People,* Leeladevi Publications, Delhi, pp. 13-43.

Ibid, pp. 13-43.

Ibid, pp. 13-43.

Chatterjee, S.N., 1984. *Tripura: A Profile,* Inter- India Publications, New Delhi, pp.1-61.

Ibid, pp. 1-61.

Ibid, pp. 1-61.

http://dhalai.gov.in/new/.html. Accessed on 3 June 2015.

DESPD. 2012. *Economic Review of Tripura (2011-12), 13th issue,* Government of Tripura, Agartala, Tripura, pp. 3-46.

Government of Tripura (GoT). 2007. *Tripura Human Development Report,* Government of Tripura, Tulika Print Communication Services, New Delhi, pp. 1-32.

Census of India. 2011. *2011 census data-Population Enumeration Data,* Office of the Registrar General and Census Commissioner, India, New Delhi. Retrieved from http://www.censusindia.gov.in/2011census/population_enumeration.html. Accessed on 4 February 2016.

GoT. 2007. *Tripura Human Development Report,* Government of Tripura, Tulika Print Communication Services, New Delhi, pp. 1-32.

Bhattacherjee, P.N. and Singh, R.G. 1995. *Tea Plantation and the Tribes of Tripura,* Tripura State Tribal-Cultural Research Institute and Museum, Government of Tripura, pp. 1-88.

Chatterjee, S.N. 1984. *Tripura: A Profile,* Inter- India Publications, New Delhi, pp. 1- 61.

Laskar, S., Dadhwal, K.S. and Prasad, R.N. 1983. Soils of Tripura and their Fertility Management, *Research Bulletin No. 23,* Division of Soil Science, ICAR Research Complex for N.E.H. Region, Shillong (Tripura Centre), pp. 1-30.

Kesari, G.K. 2011. Geology and Mineral Resources of Manipur, Mizoram, Nagaland and Tripura, *Geological Survey of India, No. 30 Part IV, Vol. 1 (Part-2),* Miscelleaneous Publication, pp. 8-10 and 78-82.

Saigal,O. 1978. *Tripura- Its History and Culture,* Concept Publishing Company, Delhi, pp. 17-27.

Chatterjee, S.N. 1984. *Tripura: A Profile,* Inter- India Publications, New Delhi, pp. 1- 61.

Ibid, pp. 1-61.

Gan-Chaudhuri, J. 1980. The Perennially Charming Tripura, in J. Gan-Chaudhuri (ed) *Tripura: The Land and its People,* Leeladevi Publications, Delhi, pp. 1- 12.

Laskar, S., Dadhwal, K.S. and Prasad, R.N. 1983. Soils of Tripura and their Fertility Management, *Research Bulletin No. 23,* Division of Soil Science, ICAR Research Complex for N.E.H. Region, Shillong (Tripura Centre), pp. 1-30.

Chatterjee, S.N. 1984. *Tripura: A Profile,* Inter- India Publications, New Delhi, pp. 1- 61.

Kesari, G.K. 2011. Geology and Mineral Resources of Manipur, Mizoram, Nagaland and Tripura, *Geological Survey of India, No. 30 Part IV, Vol. 1 (Part-2),* Miscelleaneous Publication, pp. 8-10 and 78-82.

Chatterjee, S.N. 1984. *Tripura: A Profile,* Inter- India Publications, New Delhi, pp. 1- 61.

Kesari, G.K. 2011. Geology and Mineral Resources of Manipur, Mizoram, Nagaland and Tripura, *Geological Survey of India, No. 30 Part IV, Vol. 1 (Part-2),* Miscelleaneous Publication, pp. 8-10 and 78-82.

DESPD. 2016. *Economic Review of Tripura (2015-16), 17th issue,* Government of Tripura, Agartala, Tripura, pp. 90-104.

Chatterjee, S.N. 1984. *Tripura: A Profile,* Inter- India Publications, New Delhi, pp. 1- 61.

Ibid, pp. 1-61.

Ibid, pp. 1-61.

Ibid, pp. 1-61.

Ibid, pp. 1-61.

Laskar, S., Dadhwal, K.S. and Prasad, R.N. 1983. Soils of Tripura and their Fertility Management, *Research Bulletin No. 23,* Division of Soil Science, ICAR Research Complex for N.E.H. Region, Shillong (Tripura Centre), pp. 1-30.

Chatterjee, S.N. 1984. *Tripura: A Profile,* Inter- India Publications, New Delhi, pp. 1- 61.

Saigal,O. 1978. *Tripura- Its History and Culture*, Concept Publishing Company, Delhi, pp. 17-27.

Chatterjee, S.N. 1984. *Tripura: A Profile*, Inter- India Publications, New Delhi, pp. 1- 61.

Ibid, pp. 1-61.

Saigal,O. 1978. *Tripura- Its History and Culture*, Concept Publishing Company, Delhi, pp. 17-27.

Chatterjee, S.N. 1984. *Tripura: A Profile*, Inter- India Publications, New Delhi, pp. 1- 61.

DESPD. 2016. *Economic Review of Tripura (2015-16), 17th issue*, Government of Tripura, Agartala, Tripura, pp. 129-143.

Saigal,O. 1978. *Tripura- Its History and Culture*, Concept Publishing Company, Delhi, pp. 17-27.

Chatterjee, S.N. 1984. *Tripura: A Profile*, Inter- India Publications, New Delhi, pp. 1- 61.

Bareh, H.M. 2001. *Encyclopaedia of North-East India-Volume viii-Tripura*, Mittal Publications, New Delhi, pp. 1-279.

Debbarma, S. 2009. Refugee Rehabilitation and Land Alienation in Tripura, in W. Fernandes and S. Barbora (ed.) *Land, People and Politics: Contest over Tribal Land in Northeast India*, North-Eastern Social Research Centre, Guwahati and International Work Group for Indigenous Affiars (IWGIA), Denmark, pp. 113- 127.

Bhaumik, S. 2015. Tripura: Counter-insurgency and Governance, in R. Samaddar (ed.) *Government of Peace: Social Governance, Security and the Problematic of Peace*, Ashgate Publishing Ltd., Wey Court East, England, pp. 79-106.

DESPD. 2016. *Economic Review of Tripura (2015-16), 17th issue*, Government of Tripura, Agartala, Tripura, pp. 3-40.

Bhattacharjee, S.R. 1989. *Tribal Insurgency in Tripura: A Study in Exploration Causes*, Inter- India Publications, New Delhi, pp. 23-28.

Bareh, H.M. 2001. *Encyclopaedia of North-East India-Volume viii-Tripura*, Mittal Publications, New Delhi, pp. 1-279.

DESPD. 2016. *Economic Review of Tripura (2015-16), 17th issue*, Government of Tripura, Agartala, Tripura, pp. 3-40.

Debbarma, S. 2009. Refugee Rehabilitation and Land Alienation in Tripura, in W. Fernandes and S. Barbora (ed.) *Land, People and Politics: Contest over Tribal Land in Northeast India*, North-Eastern Social Research Centre, Guwahati and International Work Group for Indigenous Affiars (IWGIA), Denmark, pp. 113- 127.

Bareh, H.M. 2001. *Encyclopaedia of North-East India-Volume viii-Tripura*, Mittal Publications, New Delhi, pp. 1-279.

Bhattacharjee, S.R. 1989. *Tribal Insurgency in Tripura: A Study in Exploration Causes,* Inter- India Publications, New Delhi, pp. 23-28.

Tripura, B. 2013. Tribal Question in Tripura: Dialogue Between Its Past and Present,

Journal of Tribal Intellectual Collective India, Vol. 1 Issue 2 No 3, pp. 33-50. DESPD. 2016. *Economic Review of Tripura (2015-16), 17th issue,* Government of

Tripura, Agartala, Tripura, pp. 3-40.

Census of India. 2011. *2011 census data-Population Enumeration Data,* Office of the Registrar General and Census Commissioner, India, New Delhi. Retrieved from http://www.censusindia.gov.in/2011census/population_enumeration.html Accessed 4 February, 2016.

Debbarma, S. 2009. Refugee Rehabilitation and Land Alienation in Tripura, in W. Fernandes and S. Barbora (ed.) *Land, People and Politics: Contest over Tribal Land in Northeast India,* North-Eastern Social Research Centre, Guwahati and International Work Group for Indigenous Affiars (IWGIA), Denmark, pp. 113- 127.

Bhatt, S.C. and Bhargava, G.K. 2006. *Land and People of Indian state and Union territories (In 36 Volumes)- Tripura, Volume- 26,* Kalpaz Publications, Delhi, pp. 17-102.

Gupta, M.D. 2007. Planning for the Hill Economy of Tripura, in A. Banerjee, B. Kar (ed.) *Economic Planning and Development of North-Eastern States,* Kanishka Publishers, New Delhi, pp. 78-79.

Census of India. 2011. *2011 census data-Population Enumeration Data,* Office of the Registrar General and Census Commissioner, India, New Delhi. Retrieved from http://www.censusindia.gov.in/2011census/population_enumeration.html. Accessed 4 February 2016.

Ibid.

DESPD. 2016. *Economic Review of Tripura (2015-16), 17th issue,* Government of Tripura, Agartala, Tripura, pp. 3-151.

Ibid, pp. 3-151.

Ibid, pp. 3-151.

Ibid, pp. 3-151.

Ibid, pp. 3-151.

Bareh, H.M. 2001. *Encyclopaedia of North-East India-Volume viii-Tripura,* Mittal Publications, New Delhi, pp. 1-279.

DESPD. 2016. *Economic Review of Tripura (2015-16), 17th issue,* Government of

Tripura, Agartala, Tripura, pp. 3-151.

Ibid, pp. 3-151.

Ibid, pp. 3-151.

Ibid, pp. 3-151.

Bhaumik, S. 2015. Tripura: Counter-insurgency and Governance, in R. Samaddar (ed.) *Government of Peace: Social Governance, Security and the Problematic of Peace*, Ashgate Publishing Ltd., Wey Court East, England, pp. 79-106.

Bareh, H.M. 2001. *Encyclopaedia of North-East India-Volume viii-Tripura,* Mittal Publications, New Delhi, pp. 1-279.

3
Spatial Pattern of Growth and Distribution of Rubber Plantation in Tripura

3.1 Introduction

Before we look into the area under rubber plantation and the spatial pattern of growth and distribution of rubber plantation in Tripura, it becomes imperative to have a general understanding of how rubber tree, a native of South America has diffused over space and how it was introduced as a plantation crop in India in general and in the state of Tripura in particular.

Natural rubber (NR) is obtained from over twenty thousand species of plants spread over approximately eighty families of flowering plants, geographically confined mainly within the tropical and subtropical climatic zones. The six principal families from where NR is obtained are *Euphorbiaceae, Moraceae, Asclepiadaceae, Apocynaceae, Sapotaceae* and *Composite* (Polhamus, 1962). However, Compagnon (1986, cited by Priyadarshan, 2011) states that there are more than seven thousand five hundred plant species that produces NR; which are confined to three hundred genera belonging to seven plant families i.e. *Euphorbiaceae, Moraceae, Asclepiadaceae, Apocynaceae, Sapotaceae, Papaveraceae* and *Asteraceae* (Archer and Audley, 1973, Heywood, 1978, Backhaus, 1985, Lewinsohn, 1991, John, 1992, Cornish et al., 1993, cited by Priyadarshan, 2011).

Since ancient time the use of NR has been known. Schurer (1957, cited by Polhamus, 1962) has discussed about the use of rubber items in Peru dating back to Mayan civilization. The ancient Mexican civilization also refers to use of NR derived from the local trees of *Castilla* species. Serier (1993, cited by Priyadarshan, 2011) has also indicates that there was use of NR mentioned derived from *Castilla* species in the ancient religious documents from Mexico dating back to 600 C.E. NR was also extracted from genus *Hevea* species mainly by the natives of South America; however with the depletion and disappearance of the primeval strands of *Castilla* species, the *Hevea* species slowly started gaining popularity among the native tappers as its latex was relatively easy to coagulate (Polhamus, 1962).

Studies (Dijkman, 1951, Baker, C.S.L, 1996, cited by Priyadarshan 2011) suggest that wild rubber trees were first discovered by Christopher Columbus during his voyage to Amazon valley from 1493 to 1496; and later by the Spanish explorers in the 15th and 16th centuries (Ochse, Soule, Dijkman and Wehlburg, 1961).

Commercial importance of NR in Europe dates back to around late 18th century when an astronomer de la Condamine sent the samples of a mysterious elastic substance (he named it 'caoutchouc' generally a phonetic rendition of the Indian name) to France from Peru in 1736 (Ochse et al., 1961; Kochhar, 1981; Dijkman, 1951, Baker, C.S.L, 1996, cited by Priyadarshan, 2011). Later, with the invention of the process of vulcanization (the application of heat to mixtures of sulphur and crude rubber which can be manufactured into various items like raincoats, boots etc) in 1839 by Charles Goodyear in United States (Ochse et al., 1961; Cobley, 1976; Kochhar, 1981) which was further improved by Hancook in England, revolutionized rubber industry overnight (Ochse et al., 1961). The rubber industry gained its peak when Dunlop reinvented the pneumatic tyre in 1888 for bicycles in England (Ochse et al., 1961; Kochhar, 1981). Soon Michelin from France and B.F. Goodrich from United States of America used NR for manufacturing tyres for bicycles and automobiles (Ochse et al., 1961). However, as early as 1860, Collins (an Englishman) warned his government of the possible difficulties that might arise in future when the exportation of wild Para rubber from Brazil would fail to meet the steadily increasing demand for NR to feed this industry. As the demand for NR increased, Hooker and Markham of the British Indian Office started investigating about the possibility of introducing rubber trees in India so as to supply raw material for the British rubber industry (Ochse et al., 1961; Kochhar, 1981). In the 19th century, impressed by the importance of NR, plantation of *Castilla* species were initially introduced in Mexico, *Hevea* species was introduced in Brazil and towards the end of the century *Funtumia elastic* in Africa while *Ficus elastic* was planted in India by the Colonial powers of Europe in their respective colonies. The rubber plantation failed in both Brazil and Mexico (Polhamus, 1962) but later *Hevea* genus was introduced in India and survived thus began the diffusion of *Hevea* rubber trees in India.

The genus *Hevea* comprises of ten plant species belonging to the *Euphorbiacea* family. This plant species has been one of the widely distributed for obtaining NR (Peter, 2002). Of these species, *Hevea Brasiliensis* (Para rubber tree or rubber tree) was widely grown as plantation to be commercially exploited as it is of great economic importance producing good quality latex for NR (Mors and Rizzini, 1966; Kochhar, 1981; Opeke, 1982; Peter, 2002; Ranganath, 2004; Gleason, 2005; Raj et al., 2012). Geographically, *Hevea Brasiliensis* is an indigenous plant of tropical regions of South America found abundantly in the Amazon basin of Brazil and adjoining areas (Ochse et al., 1961; Polhamus, 1962; Mors and Rizzini, 1966; Peter, 2002; Roy, 2005; Priyadarshan, 2011). This NR trees has been widely cultivated as plantation in Africa and America as well as in Asia by the European Colonial powers (Peter, 2002). Consequently large rubber plantation came up in Malaysia, Thailand, Indonesia, China, India, Sri Lanka, Nigeria, Liberia, Ivory Cost, Vietnam, Cambodia, Philippines, Brazil, Burma, and Mexico etc by the end of the 19th century (Borthakur, 1992) as it has higher adaptability to a wide range of agro-climatic conditions, resulting to the success and development of rubber plantation in these countries. In recent history, rubber tree has been termed as the most far reaching and successful domestication among the cash crop species (Raj et al., 2012; Rubber Board) with plantation area covering over 9.3 million hectares (ha) and out of the total area under rubber plantation 95 percent of area is concentrated in Asia (Rubber Board). The introduction of rubber plantation in most of the tropical areas of the world is a result of the European colonisation (George, Haridasan and Sreekumar, 1988) where forests were cleared to set up rubber plantation to feed the demand of NR of the colonial powers; the motive here was mainly extraction of raw material to feed the rubber industries located in Europe, leading to pauperisation and exploitation of the colonies.

3.1.1 Rubber plantation in India (Colonial phase)

In India, initially rubber plants was raised from *Hevea* seeds that were transported through present Kolkata port in 1873 (erstwhile Calcutta) and planted in adjoining areas (Kochhar, 1981; Thomas, 2002, cited by Priyadarshan, 2011), however this venture was not successful (Kochhar, 1981). During the same year, rubber trees was also introduced in South India from Ceylon (now Sri Lanka) when 28 *Hevea* plants was planted in the Nilambur Valley of present state of Kerala (Haridasan and Nair, 1980, cited in Priyadarshan, 2011). Later, around 1880-1882, rubber trees were planted on an experimental basis in different parts of South India and the Andaman islands (Priyadarshan, 2003a, cited by Priyadarshan, 2011). Around 1876 to 1895 rubber cultivation started mainly in the hill slopes of South Kerala (Tiwari, 2008; Roy, 2005).

In Northeast India there has not been any effort to set up rubber trees on a commercial scale before 1900, although indigenous rubber trees (*Ficus Elastica*) were found here, which were tapped on a large-scale in Assam before the introduction of Para rubber (*Hevea Brasiliensis*); and studies suggest that during 1880-1890 the average production of latex was between 200 to 400 tonnes annually from the Assam province of British India (Haridasan, 1975, cited by George K. et al., 1988).

It was in 1902 that the British colonial powers introduced rubber trees and set up plantation in a large scale by clearing forests initially in Thattakad in the erstwhile Travancore state (Polhamus, 1962; Sinha, 2010) and later in the present states of Karnataka, Tamil Nadu as well as other parts of Kerala with a motive of commercially exploiting the region to supply the raw material for the rubber industry set up in Europe (Polhamus, 1962; Raghavan, 2009). The main factors for the successful introduction of rubber plantation here by the British were the favourable agro-climatic conditions and the availability of cheap land and labour forces (Polhamus, 1962; George K. et al., 1988). During the same year i.e. in 1902, Periyar syndicate was formed by Murphy, Hunter and Nicoll and Ross in Travancore for planting Para rubber trees and by 1914 Para rubber trees ousted other species of rubber trees as this variety was ideally suited to the South Indian agro-climatic conditions leading to higher productivity and quality of latex. Travancore emerged as the leading centre of rubber plantation although rubber plantation was first introduced in the erstwhile Cochin state in 1905. As rubber plantation proved to be a profitable venture some of the native Indians also set up plantation along with the Britishers (George K. et al., 1988).

3.1.2 Rubber plantation after Indian Independence

After India independence, the dynamic growth of area under rubber plantation was propelled mainly by the Indian government. The government of India initially followed two main schemes concerning rubber industry, *firstly* to enhance production for meeting the growing internal demand by increasing the productivity in existing areas and extension of cultivation to new areas, and *secondly* to ensure a remunerative price as an incentive to the growers so as to achieve the goal of self-sufficiency in production. Consequently, New Planting Subsidy Scheme was introduced in 1979 mainly for increasing NR production through extensive cultivation of rubber trees which was a success. This scheme was succeeded by an integrated new scheme entitled Rubber Plantation Development Scheme (Phase I: 1980-84) where the subsidy schemes of new plantation and re-plantation merged. The achievement of physical targets set up during Phase I led to upgrading of the scheme to continue in Phase II (1985-89), however only growers having an area up to 5 ha were eligible for the subsidy scheme in the traditional areas of rubber plantation. However, in the non- traditional areas where rubber was introduced after independence, all categories of growers were eligible for the subsidy scheme (George K. et al., 1988). Subsequently, Northeast India emerged as an important non-traditional centre for rubber production.

In the traditional areas of rubber plantation i.e. Kerala, initially had concentration of rubber plantation in the districts of Idukki, Kottayam and Pathanamtitta but this spread further to the adjoining areas as the profit margin of this crop increased with increasing demand whereas crops like rice and coconut fell, consequently farmers shifted to rubber which enhanced the growth of rubber plantation extensively (Rajasenan, 2010). At present, Kerala and Kanyakumari district of Tamil Nadu (the traditional rubber growing regions of India) accounts for about 90 percent of the total NR production (DESPD, 2016).

India at present is the fifth largest producer of NR (DESPD, 2016). The increasing demand for NR and the saturation limit of areas for further expansion of rubber trees in the traditional areas has created a situation where the Rubber Board of India along with the government of India has ventured into new areas (non-traditional regions). Consequently, some of the Northeast states of India are found suitable for setting up of rubber plantation (Vinod et al., 1996; Bhowmik, 2006; Vinod, Meenattoor, Reddy, Priyadarshan and Chaudhuri, 2010; Raj et al., 2012).

Around late 1960s the Rubber Board along with the government of India took up initiatives to set up rubber plantation in the Northeast states of India not only to increase rubber production in the country through extensive cultivation of rubber plantation in non-traditional regions so as to achieve self-sufficiency in NR and import substitution in all important spheres of rubber production but also to rehabilitate the economically and socially marginalized groups found in the region who are into *jhum* cultivation practices and lacking in socio-economic development. Consequently, the government along with the Rubber Board took up the initiative to integrate marginalized *jhumias* (both ST and SC population) into the mainstream of development through policy intervention (Joseph, George and Dey, 2010; Sharma, Joseph, George K and Dey, 2011). The surveyed reports indicated that the Northeast India had both geographical and agro-climatic conditions favourable for rubber plantation (Borthakur, 1992). Further the National Bureau of Soil Survey and Land Use Planning (NBSS & LUP) identified that 4, 50,000 ha of land in Northeast India having high potential for rubber plantation (Bahuguna, 2005, cited by Bhowmik, 2006). Of the Northeast states of India, Tripura was identified to have a large potential area ideal for setting up of rubber plantation (Vinod et al., 1996). Consequently, rubber plantation was introduced for the first time in undivided state of Assam in 1955 and here the rationale behind such plantation was very different from the colonial power. From 1963 to 1965, the Rubber Board took up various exploratory surveys in introducing rubber trees in the undivided state of Assam and Tripura which was found to be socio-economically viable (Borthakur, 1992). Encouraged by the trail plantation, the Rubber Board started rubber plantation for commercial purpose since late 1980s in the Northeast India (Viswanathan, 2006, cited by Viswanathan and Bhowmik, 2014). In Assam and Tripura initially the Public Sector Corporations set up rubber plantation, later this was introduced in other states of Northeast India such as in Meghalaya and Mizoram by the State Soil Conservation Department, in Manipur and Nagaland by State Forest Department (Sarkar, 1994). Arunachal Pradesh also came within the ambit of rubber plantation (Gupta, 2000). It must be mentioned here that besides traditional rubber growing region in the country i.e. Kerala and Kanyakumari district of Tamil Nadu, rubber plantation has also been introduced and distributed in the non-traditional area of the country i.e. Tripura, Assam, Meghalaya, Mizoram, Nagaland, Manipur, Arunachal Pradesh, Karnataka, Goa, Maharashtra, Andhra Pradesh, Orissa, West Bengal and Andaman & Nicobar Islands (Centre for Development Studies) which is shown in figure 3.1.

3.2 Rubber Plantation in Tripura

Rubber plantation in Tripura was initiated by the state government way back in the early 1960s mainly as a programme for afforestion and soil conservation, but soon it became a measure to rehabilitate the marginalized ST *jhumias* and other landless marginalized population to alternative livelihood options and winning over the *jhumias* to settled agriculture with a rationale of enhancing their socio-economic conditions.

Rubber plantation was set up in 1963 on a trial basis by the Forest and Soil Conservation Department (Vinod et al., 1996) basically as afforestation programme for soil conservation measure in Patichari and Manu hill ranges of South Tripura and North Tripura districts of the state (Bhowmik, 2006; Viswanathan and Bhowmik, 2014) on the degraded forestland caused due to *jhum* cultivation and excessive felling of forest trees. The success of the trial plantation motivated the Forest Department for more extensive cultivation of rubber trees (Chakravarti, 1986; Bhowmik, 2006; Raj et al., 2012) but rubber trees being a new commercial crop had few takers as the tribes out rightly rejected this new venture. However, with the rubber trees that was introduced in 1963 maturing in the early 1970s and by 1971-72 commercial tapping and processing of rubber started, the local tribes soon realized the economic importance and ecological benefits of such plantation (Bhowmik, 2006; Viswanathan and Bhowmik, 2014). Consequently, the tribes now are open to taking up such ventures. Encouraged by the success of rubber

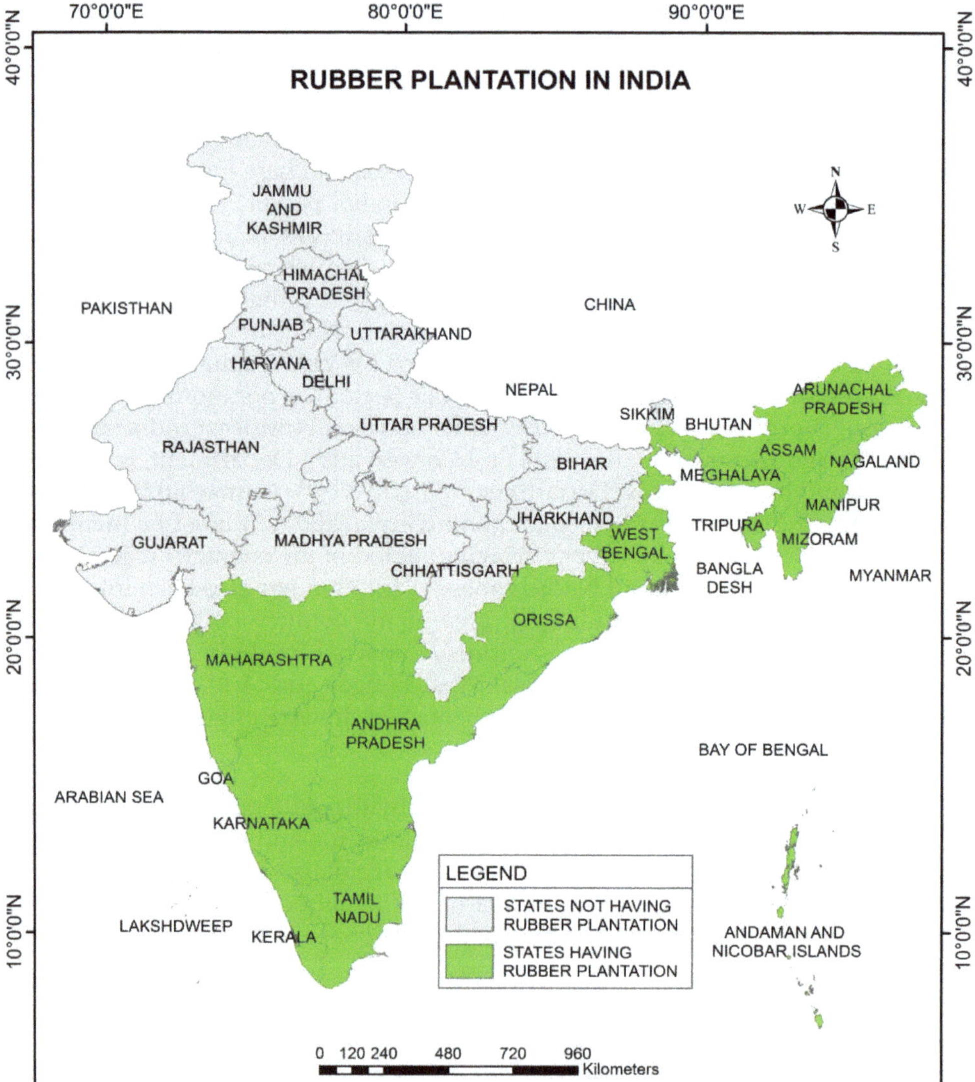

Fig. 3.1 : State wise distribution of rubber plantation in India.

plantation in not only checking deforestation and soil erosion but also contributing towards the state revenue, the state government soon realized that rubber plantation has potential for economically rehabilitating the marginalized *jhumias* and landless population to settled agriculture in order to enhance their socio-economic development (Viswanathan and Bhowmik, 2014). Subsequently, the first model to explore the possibilities of resettlement of landless tribal, *jhumias* and other marginalised ST communities through rubber plantation was first implemented at Warrangbari village in West Tripura district by the TFDPCL in 1976 (Bhowmik, 2006; Viswanathan and Bhowmik, 2014). Along with the TFDPCL, the state government also established other agencies like TRPCL in 1983 and also encouraged the schedule tribal/caste welfare department of the state in collaboration with the Rubber Board to introduced the TBPP/BPS in 1992; further the office of TTAADC and SDM/BDO were also asked to promote rubber plantation and initiate rehabilitation programme from 1998 onwards (Bhowmik, 2006; Joseph, 2010; Viswanathan and Bhowmik, 2014). The rationale behind the expansion of rubber plantation by these agencies along with assistance from government officials like BDO/SDM etc are basically to rehabilitate the *jhumias* and other landless families mainly belonging to ST communities from *jhum* to settled agriculture and provide them with alternative livelihood options which can in the long run help in enhancing their socio-economic conditions (Sarkar, 1986; Bhowmik, 2006; Raghavan, 2009; Choudhury, 2009; Viswanathan and Bhowmik, 2014).

Study undertaken by NBSS and LUP approved by the Rubber Board and government of Tripura, supervised by government of Tripura with special reference to major crops of the state and rubber plantation indicated that approximately 1 lakh ha of land is suitable for cultivation of rubber trees in the state of Tripura (Bhattacharyya, Sehgal abd Sarkar, 1996; Bahuguna, 2005, cited by Bhowmik, 2006; Raghavan, 2009; Bhattacharjee, 2012). As per Joseph et al., 2009; 2010; 2012, cited by Sharma, George K and Dey (2014) Tripura is the only state among the non-traditional rubber growing states where potential area for rubber plantation has been utilized to the highest level. Consequently, Tripura is declared the 'Second Rubber Capital of India' next to Kerala by the Rubber Board (Viswanathan and Bhowmik, 2014; Rubber Board, 2016; DESPD, 2016).

The expansion of area under rubber plantation in the state was taken up by both the public sectors (government undertaking) private sector and individual owners. While, the private sector and individual owned rubber plantation are mainly for profit proposition, the public sector owned rubber plantation-based rehabilitation programme under taken by different agencies are more a part of the development programme (Bhowmik, 2006).

3.2.1 Rubber Plantation as Rehabilitation Programme

Under the rubber plantation based rehabilitation programme, as already mentioned the beneficiary of this scheme enjoys land occupancy rights over 1 ha of land per family. The beneficiary also derives income by selling the rubber latex and rubber sheets after a gestation period of seven years when rubber trees mature and are ready for latex production. Consequently, the beneficiaries of such plantations are employed as a wage labourer in their own plantation for the initial seven years when land development and other production activities are taken up like inter-cropping of fruits like banana, pineapple etc for the market, this encourages alternative livelihood practices. Studies suggest that the resettlement programme carried out by formation of co-operatives are highly successful as they combine a number of beneficiaries as a unit so that they can enjoy better income in the production process (Sinha, 2007, cited by Viswanathan and Bhowmik, 2014).

As per DESPD (2008) the rehabilitation programme for permanent settlement of the ST *jhumias* through rubber plantation has proved to be extremely successful. Devi (2010) states that suitability of the terrain and acceptability of the economic value of rubber by the people has made rubber plantation as one of the thrust area for development leading to alternative livelihood practices and socio-economic development of the marginalized population in Tripura. Among the Northeast states of India, Tripura has been the most successful state in weaning away the native tribes from the practice of *jhum* cultivation to settled cultivation providing livelihood security through rubber plantation for the past four decades (Sharma et al., 2011). The rubber plantation based rehabilitation programme has been noted as one of the most promising prospects for the ST *jhumias* and landless people resettlement programme in the state as it has not only improved the socio-economic and welfare of the beneficiaries by providing gainful livelihood but also provides employment opportunities for the unemployed populace (Chakravarti, 1986; Bhowmik; 2006; Maithani, 2005; Raghavan, 2009; Raj et al., 2012). The rubber plantation sector has been considered an important source of economic dynamism for more than 57,000 households as well as for the state economy (DESPD, 2015; Bhowmik and Viswanathan, 2015).

3.3 Emerging pattern of increase of area under rubber plantation in Tripura

As mentioned rubber plantation was introduced in Tripura during 1963, but as there is no yearly data available regarding area under rubber plantation, it is from 1972-73 onwards that an attempt has been made to assess the area under rubber plantation in Tripura.

3.3.1 Share of area under rubber plantation of Tripura to the total area under rubber plantation of the country

As depicted in table 3.1, the share of area under rubber plantation in Tripura during 1972-73 was only 0.05 percent of the total area under rubber plantation in India which at present has increased to 9.17 percent (2015-16) as a result of continuous increase in the expansion of area under rubber plantation in the state, except in the year of 2013-14 where the increasing trend of share was found slightly lower than the previous year (figure 3.2) and the reason for this can be attributed to the increase in area under rubber plantation in other rubber growing states. From 1972-73 to 2015-16 (about four decades), the average share of area under rubber plantation of the state to the total area under rubber plantation of India is to the tune of 3.88 percent.

Table 3.1: Share of area under rubber plantation of Tripura to the total area under rubber plantation of India (1972-73 to 2015-16)

Year	Total area under rubber plantation (in ha) in India	Total area under rubber plantation (in ha) in Tripura	Share (in %) of area under rubber plantation of Tripura to the total area under rubber plantation of India
1972-73	2,23,465	103	0.05
1973-74	2,27,317	223	0.10
1974-75	2,31,452	307	0.13
1975-76	2,35,876	417	0.18
1976-77	2,40,593	574	0.24
1977-78	2,45,200	878	0.36
1978-79	2,53,279	1,306	0.52
1979-80	2,65,211	1,989	0.75
1980-81	2,84,166	2,746	0.97
1981-82	3,01,924	3,590	1.19
1982-83	3,21,495	4,578	1.42
1983-84	3,39,848	5,338	1.57
1984-85	3,61,960	6,545	1.81
1985-86	3,82,831	7,742	2.02
1986-87	4,02,329	10,085	2.51
1987-88	4,21,512	12,670	3.01
1988-89	4,40,584	14,326	3.25
1989-90	4,60,341	16,620	3.61
1990-91	4,75,083	17,320	3.65
1991-92	4,88,514	17,860	3.66
1992-93	4,99,374	18,250	3.65
1993-94	5,08,420	18,670	3.67
1994-95	5,15,572	19,252	3.73
1995-96	5,24,075	20,761	3.96
1996-97	5,33,246	21,982	4.12
1997-98	5,44,534	22,582	4.15
1998-99	5,53,041	24,120	4.36
1999-00	5,58,584	25,380	4.54
2000-01	5,62,670	26,495	4.71
2001-02	5,66,555	27,947	4.93
2002-03	5,69,667	28,853	5.06
2003-04	5,75,980	30,770	5.34
2004-05	5,84,090	32,065	5.49

Year	Total area under rubber plantation (in ha) in India	Total area under rubber plantation (in ha) in Tripura	Share (in %) of area under rubber plantation of Tripura to the total area under rubber plantation of India
2005-06	5,97,610	34,189	5.72
2006-07	6,15,200	35,760	5.81
2007-08	6,35,400	39,670	6.24
2008-09	6,61,980	46,588	7.04
2009-10	6,86,515	51,292	7.47
2010-11	7,11,560	53,669	7.54
2011-12	7,34,780	57,621	7.84
2012-13	7,57,520	61,231	8.08
2013-14	7,78,400	62,529	8.03
2014-15	7,95,135	70,296	8.84
2015-16	8,10,800	74,335	9.17

Source: Centre for Development Studies; Rubber Board, 2016 (share of area under rubber plantation computed by the researcher).

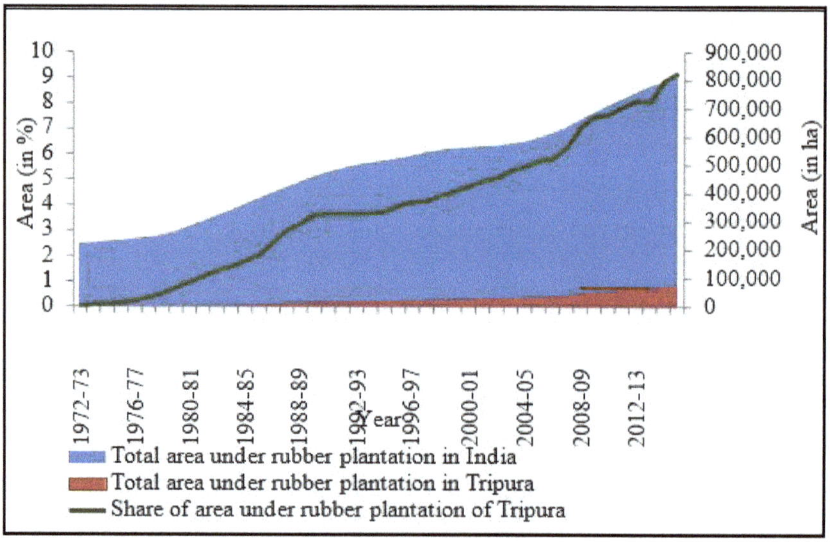

Fig. 3.2: Share of area under rubber plantation of the state to the total area under rubber plantation of India (1972-73 to 2015-16).

A further analysis suggest that marked phases can be identified in regards to share of area under rubber plantation of the state in context to the total area under rubber plantation of India. It is observed that in the first quarter (1972-73 to 1982-83) the share of area under rubber plantation of the state suggested a marginal increase where the average share is as low as 0.54 ha. In the next quarter (1984-1994) the share of area under rubber of the state suggested a slow increase as the average share is 2.95 ha. In the third quarter (1995-2005) the share of area under rubber plantation of the state suggested further increase than the previous quarter with the average share of 4.58 ha. In the last quarter (2006-2016) the share of area under rubber plantation of the state suggested an increase in the share than the previous quarter with an average of 7.43 ha. Consequently, it is observed that Tripura is emerging as an important rubber growing area of India where the share in the expansion of area under rubber plantation is increasing over the years.

3.3.2 Expansion of area under rubber plantation in Tripura

Area under rubber plantation in Tripura during its inception (1963-64) was only 8 ha (Das, 2014) which during (2015-16) has increased to 74,335 ha (appendix 1) suggesting that expansion of area under rubber plantation in the state of Tripura has been tremendous with an overall growth rate as high as 9, 29,087.5 percent within a span of about five decades (1963-64 to 2015-16). Area under rubber plantation in Tripura from 1972-73 to 2015-16 is shown in appendix 1.

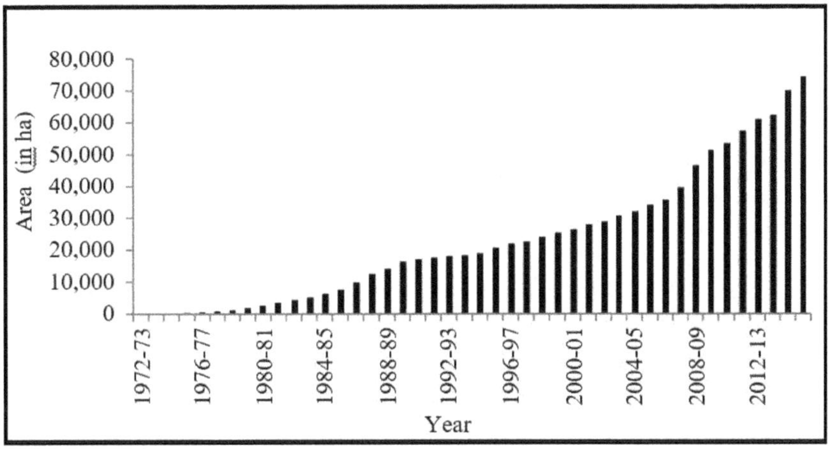

Fig. 3.3: Total area under rubber plantation in Tripura (1972-73 to 2015-16). Source: Rubber Board, 2016; Centre for Development Studies.

It is observed from figure 3.3 that the rapid expansion of area under rubber plantation in the state geared-up from the early 1980s and another major expansion of area under rubber plantation in the state was felt from 1990s onwards. This expansion of area under rubber plantation in the state are mainly due to-

i) The introduction of cash subsidy by the Rubber Board around 1980s in order to expand area under rubber plantation throughout the rubber growing states of Northeast India including Tripura (Viswanathan and Bhowmik, 2014).

ii) Increase in the price of NR due to economic reforms and succeeding industrial spurt experienced by the country as a whole and the state in general which attracted large number of private entrepreneurs especially around 1994 and 1995 as private parties invested on rubber plantations here (Viswanathan and Bhowmik, 2014).

iii) The revival of demand for NR in the international market from 2001 onwards soon after the depression in the demand of NR which was experienced around 1990s. This renewed the interest among the private and individual investors to invest on rubber plantation consequently the area under rubber plantation increased (Viswanathan and Bhowmik, 2014). Mohanakumar (2013 and 2014) also stated that the expansion of area under rubber has been rapid in Tripura due to increase in price of NR in the national/domestic and international markets.

iv) The implementation of rubber plantation-based rehabilitation programme for the marginalized section of people in the state by the state government has also played an important role in the expansion of area under rubber plantation in the state (Bhowmik, 2006; Joseph et al., 2010; Viswanathan and Bhowmik, 2014).

As already mentioned about 1 lakh ha of the state's geographical area is found suitable for rubber plantation and by 2015-16 about 74.34 percent (74,335 ha) of area earmarked for rubber plantation has come under rubber plantation indicating that around 25,665 ha (25.665 percent) of area earmarked for rubber is still available for further cultivation of rubber trees in the state. Thus, there is scope for further expansion of area under rubber plantation in the state.

1.3.3 Nature of growth of area under rubber plantation in Tripura

Area under rubber plantation in the state has been increasing continuously and to understand the trend of annual growth rate in the expansion of area under rubber plantation, the annual growth rate of area under rubber plantation in Tripura from 1972-73 to 2015-16 is shown in figure 3.4 (also see appendix 1). It is observed from figure 3.4 that area under rubber plantation in the state is continuously increasing but at a declining and fluctuating growth rate. The annual average growth rate of area under rubber plantation in the state from 1972-73 to 2015-16 is to the tune of 18.05 percent per year. The highest annual growth rate of area under rubber plantation in the state was experienced during 1973-74 having an annual growth rate of more than 100 percent (116.50 percent) while the lowest annual growth rate of area under rubber plantation was experienced during 2013-14 with just 2.12 percent growth (appendix 1).

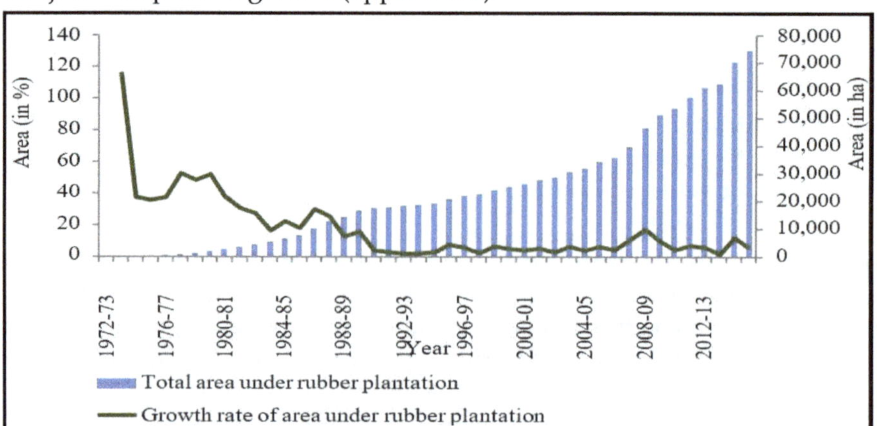

Fig. 3.4: Growth rate of area under rubber plantation in the state (19972-73 to 2015-16).

1.3.3.1 Decadal growth of area under rubber plantation in Tripura

The introduction of rubber plantation in Tripura has crossed 53 years roughly corresponding to about five decades. An assessment of the decadal growth of area under rubber plantation (table 3.2) suggests the following:-

By the end of the first decade (1963-64 to 1972-73) area under rubber plantation in the state was 103 ha suggesting a slower growth in comparison to the next decade as the compound annual growth rate (CAGR) was 28.67 percent. In this decade, the expansion of area under rubber plantation was initiated mainly by the Forest Department, government of Tripura (Mohanakumar, 2013).

By the end of the second decade (1973-74 to 1982-83) area under rubber plantation in the state was 4,475 ha having a CAGR of 35.28 percent suggesting a faster CAGR in comparison to the previous decade. The expansion of area under rubber plantation during this decade were initiated through rubber plantation-based rehabilitation programme and commercial purposes mainly implemented by the TFDPCL (Mohanakumar, 2013).

Table 3.2: Decadal CAGR of area under rubber plantation in Tripura

Decadal year (DY)	Area under rubber plantation (in ha)	CAGR of area under rubber plantation (in %)
1963-64 to 1972-73 (1st DY)	103	28.67
1973-74 to 1982-83 (2nd DY)	4,475	35.28
1983-84 to 1992-93 (3rd DY)	13,672	13.08
1993-94 to 2002-03 (4th DY)	10,603	4.45
2003-04 to 2012-13 (5th DY)	32,378	7.12

Source: Decadal CAGR computed by the researcher.

By the end of third decade (1983-84 to 1992-93) area under rubber plantation in the state was 13,672 ha with the CAGR of 13.08 percent suggesting a slower CAGR in comparison to the previous decade. In this decade, the expansion of rubber plantation was also initiated through rehabilitation programme implemented by both the TFDPCL and the TRPCL (Mohanakumar, 2013).

By the end of the forth decade (1993-94 to 2002-03) area under rubber plantation in the state was 10,603 ha with the CAGR of 4.45 percent suggesting a further slowing down in the CAGR of area under rubber plantation in comparison to the previous decade. The expansion of area under rubber plantation during this decade was mainly launched through rubber plantation-based rehabilitation programme known as Block Planning Scheme (BPS)/Block Plantation Project (BPP) integrated and implemented jointly by the Rubber Board and the Tribal Welfare Department, government of Tripura (Mohanakumar, 2013) besides the area under rubber plantation developed by the TFDPCL and the TRPCL. By the end of the fifth decade (2003-04 to 2012-13) area under rubber plantation was 32,378 ha with the CAGR of 7.12 percent suggesting a slightly higher CAGR in comparison to the previous decade. The expansion of area under rubber plantation during this decade was initiated mainly by the private sectors and individual owners.

4.3.4 Annual gain in area under rubber plantation in Tripura

As already mentioned rubber plantation was introduced in Tripura during 1963, but in the initial phase area under rubber plantation showed a marginal increase as the rubber plantation-based rehabilitation programme was not very popular and there is not much data available suggesting the rate of annual increase of area under rubber plantation. It is from 1972-73 onwards the yearly data on area under rubber plantation is available. Consequently, an attempt has been made to assess the annual gain in area under rubber plantation for a period of about four decades; further an attempt has also been made to assess the trend in the increase of area under rubber plantation from 1973-74 onwards till 2015-16 by the technique of three years moving average.

The annual gain in area under rubber plantation in Tripura from 1973-74 to 2015-16 is depicted in figure 3.5. The figure suggests that the lowest gain in area under rubber plantation in the state was during 1974-75 with a marginal gain of 84 ha while the highest gain in area under rubber plantation was during 2014-15 with 7,767 ha (appendix 1). From 1973-74 to 2015-16, the average gain in area under rubber plantation in the state is 1,726.33 ha. The standard deviation is 1,707.76 and the coefficient of variation is to the tune of 98.92. This suggests that there is a very large variation in the annual gain of area under rubber plantation in the state from 1972-73 to 2015-16.

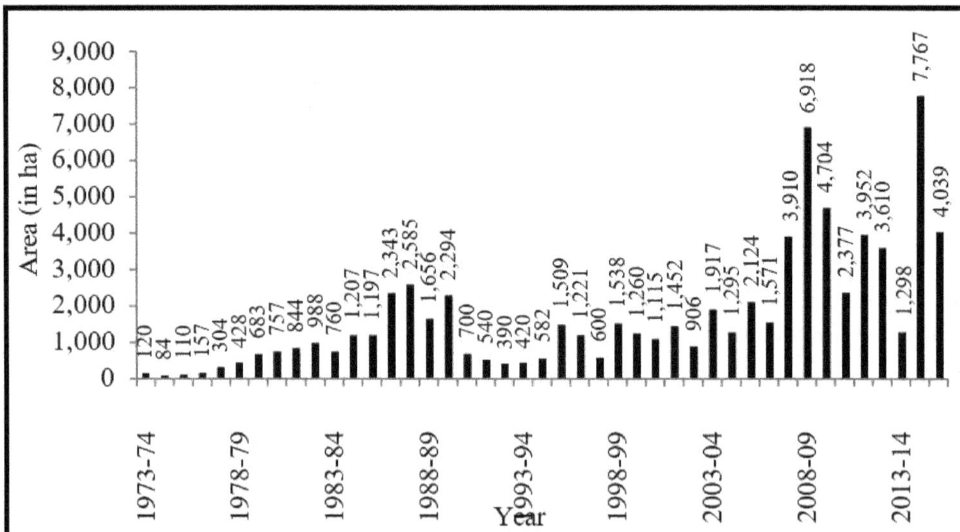

Fig. 3.5 : Annual gain in area under rubber plantation in Tripura (19973-74 to 2015- 16).

In order to understand the nature of the distribution of annual gain in area under rubber plantation in the state, skewness has been computed. Figure 3.6 suggest a very large positive skew in the distribution of annual gain in area under rubber plantation in the state as the distribution has longer/extended curve tail to the right direction and also the average mean value of the gain in area under rubber plantation (1,726.33 ha) is greater than the median value (1,221). Applying the Pearson second coefficient of skewness, the coefficient of skewness of the annual gain in area under rubber plantation in the state is to the tune of 0.89 suggesting that there is large degree of positive skewness in the distribution of the annual gain in area under rubber plantation in the state.

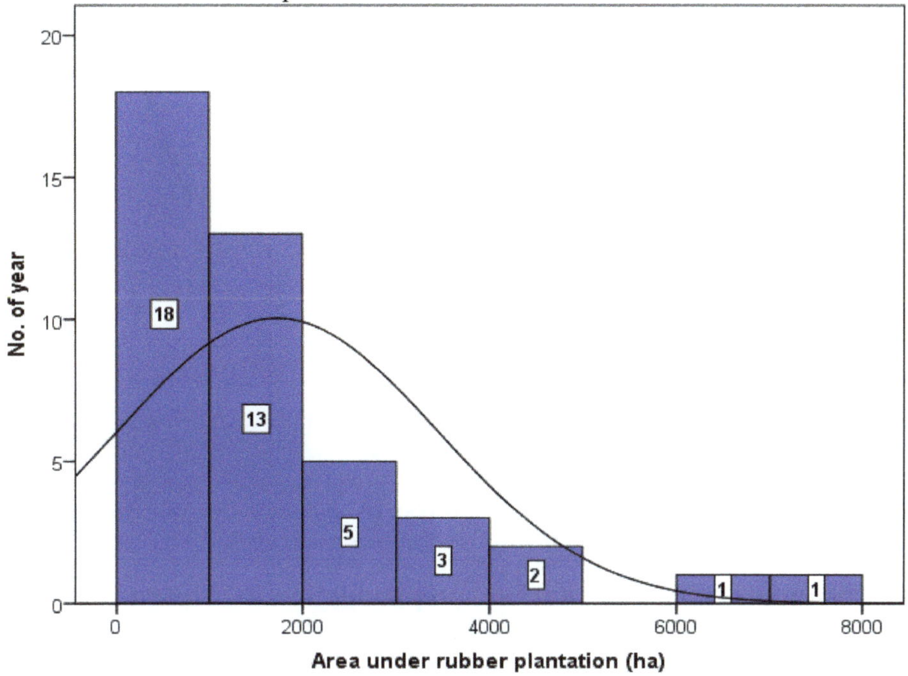

Fig. 3.6: Nature of annual gain in area under rubber plantation and number of year in the state (1972-73 to 2015-16).

It is observed from figure 3.6 that in most of the year i.e. 31 number of year (out of the 43 number of year) annual gain in area under rubber plantation in the state is below 2,000 ha and in about 10 number of year the annual gain in area under rubber plantation in the state is between 2,000-5,000 ha while only in 2 number of years the annual gain in area under rubber plantation in the state is between 6,000-8,000 ha.

The trend line (3 yearly moving average) in the annual gain in area under rubber plantation in the state is depicted in figure 3.6 (also see appendix 1). Trend line in the annual gain in area under rubber plantation indicates that from 1974-75 the annual gain in area under rubber plantation has been increasing continuously reaching its peak in 1987-88. After 1988, the annual gain in area under rubber plantation suggests a declining trend till 1992-93. From 1993-94 to 2005-06 the annual gain in area under rubber plantation suggest a fluctuating trend. After 2006, the annual gain in area under rubber plantation again started increasing continuously reaching a peak during 2008-09. Subsequently from 2009-10 the annual gain in area under rubber plantation started declining drastically till 2012-13 however after 2013, the annual gain in area under rubber plantation in the state again started increasing. The trend line suggest that there are two peaks in the annual gain in area under rubber plantation in the state i.e. one during 1987-88 and the other during 2008-09.

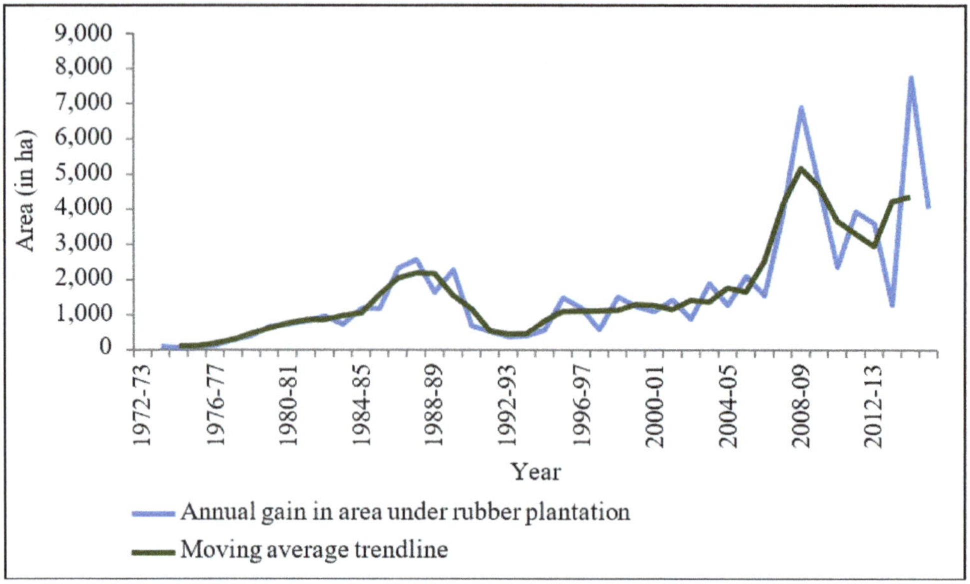

Fig. 3.7: Trend of annual gain in area under rubber plantation in the state from 1974-75 to 2015-16 (applying 3 year moving average).

1.4 Area under rubber plantation and production in Tripura

It is interesting to note that there is a positive co-relation between area under rubber plantation and the production of NR in the state which is to the tune of 0.98 suggesting a very high co-relation between increasing area under rubber plantation and the production of NR from 2006-07 to 2015-16 (table 3.3 and figure 3.8).

Table 3.3: Area under rubber plantation and its production in Tripura (2006-07 to 2015-16)

Year	Area under rubber plantation (in ha)	Production (in Mt)
2006-07	35,760	20,000
2007-08	39,670	23,087
2008-09	46,588	25,588
2009-10	51,292	26,191
2010-11	53,669	29,698
2011-12	57,621	32,332
2012-13	61,231	37,277
2013-14	62,529	39,747
2014-15	70,296	44,740
2015-16	74,335	49,943

Source: Rubber Board, cited by DESPD (2009, 2010, 2011, 2012, 2013, 2014, 2015, 2016); Rubber Board, 2016.

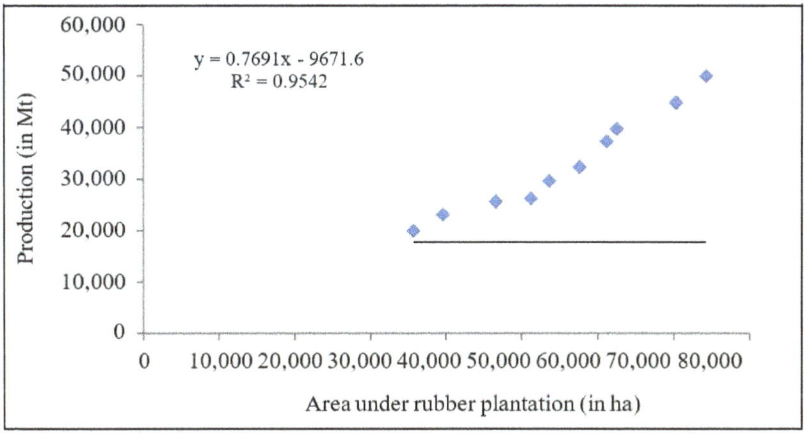

Fig. 3.8: Nature of correlation of area under rubber plantation and production in Tripura (2006-07 to 2015-16).

1.5 Distribution of area under rubber plantation in Tripura

As mentioned the district wise date regarding area under rubber plantation are available from 2008-09 onwards, consequently the district wise distribution of area under rubber plantation has been analyzed from 2008-09 to 2015-16.

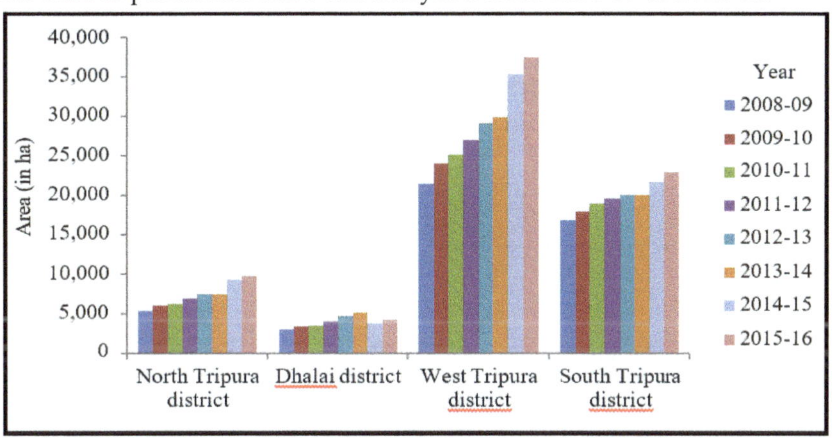

Fig. 3.9: District wise area under rubber plantation in Tripura (2008-09 to 2015- 16).

Source: Rubber Board, cited by DESPD (2009, 2010, 2011, 2012, 2013, 2014, 2015,2016); Rubber Board, 2016.

Figure 3.9 gives an idea about the district wise distribution of rubber plantation in the state of Tripura from 2008-09 to 2015-16 (also in appendix 2). The figure suggests that from 2008-09 to 2015-16 maximum area under rubber plantation is reported in West Tripura district followed by South Tripura and North Tripura districts and the least area under rubber plantation is observed in Dhalai district. It is interesting to note that it is in Dhalai district (in Manu hill ranges of the district as cited by Bhowmik, 2006; Viswanathan and Bhowmik, 2014) that the first rubber plantation was introduced on an experimental basis to check soil erosion and deforestation way back in 1963-64 yet at present this district has the lowest area under rubber plantation in spite of the fact that this district has the highest share of ST population (i.e. 55.68 % as per Census of India, 2011) in the state where most of the ST population in the district practices *jhum* cultivation and the rehabilitation programme for the *jhumias* and other marginalized people through rubber plantation has been initiated in the state by different agencies and the Rubber Board, yet the area under rubber plantation in this district is the least. It is noticed that area under rubber plantation has been increasing continuously from 2008-09 to 2015-16 in all the districts except in Dhalai district in the recent years which is clearly shown in figure 3.9.

The maximum area under rubber plantation in West Tripura and South Tripura districts can be attributed to favourable conditions like availability of land suitable for rubber plantation, better accessibility, enthusiasm among the people and policy intervention by the state administration. Rubber plantation in North Tripura and Dhalai districts has been marginal as most parts of these districts are covered by hilly terrain with forested areas (Mohanakumar, 2014); In Tripura rubber plantation is not encouraged in forested areas (TRPCL, 2012 and 2016; Raj et al., 2012). As per Ministry of Environment of Government of India in 2005, cultivation of rubber trees is totally banned in forested lands i.e. mainly reserved forest areas (Mohanakumar, 2013).

1.5.1 District wise growth rate of area under rubber plantation in Tripura

The district wise growth rate of area under rubber plantation in Tripura from 2008-09 to 2015-16 is depicted in figure 3.10 (also in appendix 3). The growth rate of area under rubber plantation of the four districts as shown in the figure suggests that the area under rubber plantation is increasing at a fluctuating growth rate. However, it is observed that during 2014-15 the growth rate of area under rubber plantation in Dhalai district is negative.

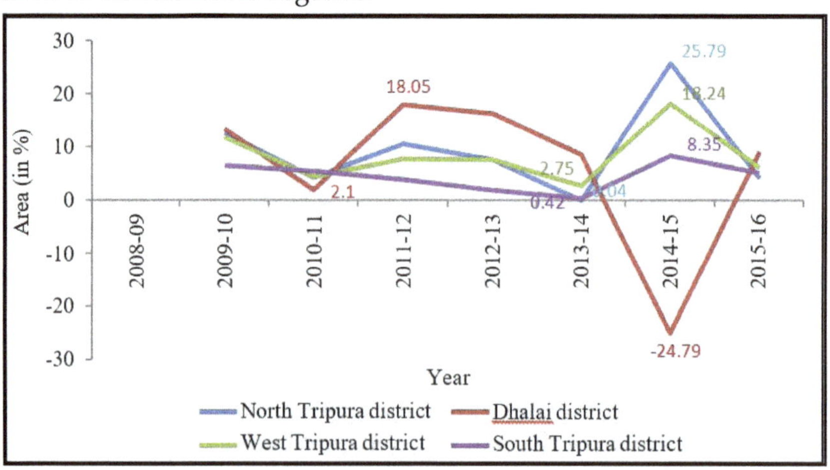

Fig. 3.10 : District wise growth rate of area under rubber plantation in Tripura (2008-09 to 2015-16).

Figure 3.10 suggest that in North Tripura district the highest annual growth rate of area under rubber plantation (25.79%) was during 2014-15 and the least annual growth rate (0.04%) was in the year 2013-14. The average annual growth rate of area under rubber plantation in the district is to the tune of 9.31 percent per annum.

In Dhalai district the highest annual growth rate of area under rubber plantation (18.05%) was in the year 2011-12 while the least annual growth rate of the area under rubber plantation was negative as reported in the year 2014-15 when the annual growth rate of area under rubber is as low as -24.79 percent suggesting shrinkage in area under rubber plantation. The average annual growth rate of area under rubber plantation in the district is 6.15 percent per annum.

In West Tripura district the highest annual growth rate of area under rubber plantation (18.24%) was reported in the year 2014-15 and the minimum annual growth rate of area under rubber plantation here was reported in 2013-14 (2.75%). The average annual growth rate of area under rubber plantation in this district is to the tune of 8.38 percent per annum.

In South Tripura district the highest annual growth rate of area under rubber plantation (8.35%) was reported in the year 2014-15 while least growth rate in area under rubber plantation was reported in 2013-14 (0.42%). The average annual growth rate of area under rubber plantation in this district is 4.53 percent per annum.

It is observed that in three districts i.e. North Tripura, West Tripura and South Tripura districts the growth rate of area under rubber plantation was the highest during 2014-15 however the growth rate in Dhalai district was noticed to be negative. This open up scope for further research.

3.5.2 District wise concentration of area under rubber plantation in Tripura

In Tripura, rubber plantation has spread throughout the four districts. With the help of concentration index, area under rubber plantation of the four districts has been computed for the year 2015-16. The concentration index (CI) score ranges from 1.77 to 0.25. Consequently, the concentration of area under rubber plantation has been categorized into high concentration, moderate concentration, low concentration and very low concentration zones.

It is observed from figure 3.11 that West Tripura district has the highest concentration of area under rubber plantation in the state with a CI score of 1.77. South Tripura district has moderate concentration of area under rubber plantation with a CI score of 1.06. North Tripura district has low concentration of area under rubber plantation with a CI score of 0.68 while Dhalai district has very low concentration of area under rubber plantation with a CI score as low as 0.25.

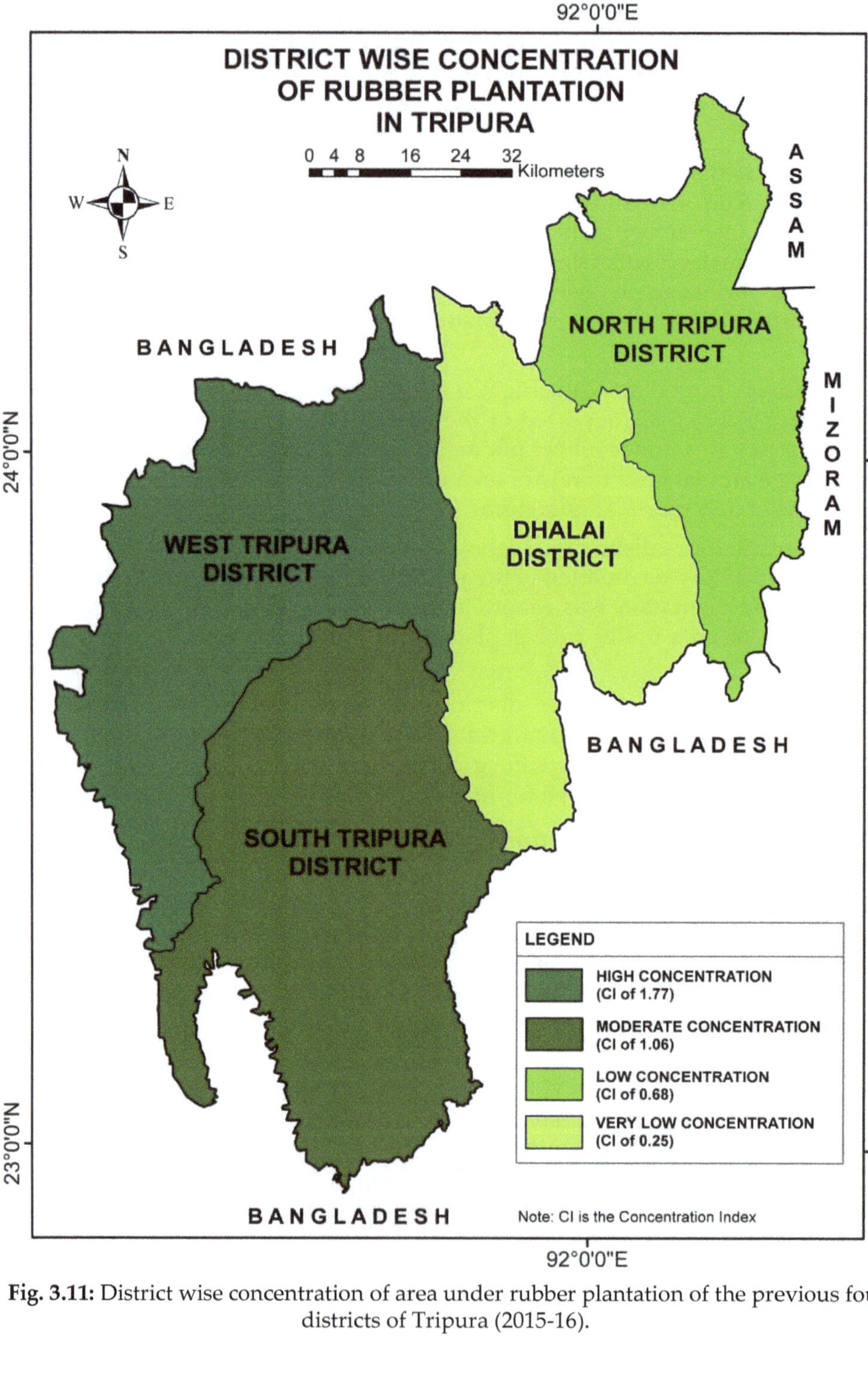

Fig. 3.11: District wise concentration of area under rubber plantation of the previous four districts of Tripura (2015-16).

At present the total districts of Tripura are eight in number (table 3.4) which has been carved out from the previous districts as the Government of Tripura has created four new districts for the purpose of effective administration in order to reach out to the people and bring in more effective inclusive development. The previous North Tripura district has been divided into North Tripura and Unakoti districts, West Tripura district has been divided into three districts viz. Khowai, West Tripura and Sipahijala districts. Previous South Tripura district has been divided into two districts viz. Gomati and South Tripura districts (DESPD, 2012). The administrative boundary of Dhalai district has not undergone any changes.

The data of area under rubber plantation for the eight districts are available only for the last two years i.e. 2014-15 and 2015-16. Consequently, an attempt has also been made to assess the share and concentration of area under rubber in all the eight districts of the state for the year of 2015-16.

Table 3.4: Area under rubber plantation in different district of Tripura (2015-16)

District	Total area under rubber plantation (in ha)	Share in % to total
North Tripura	7,134	9.60
Unokoti	2,625	3.53
Dhalai	4,230	5.69
Khowai	3,008	4.05
West Tripura	17,625	23.71
Sepahijala	16,846	22.66
Gomati	7,416	9.98
South Tripura	15,451	20.79
Total	74,335	100

Source: Rubber Board, 2016.

Table 3.4 suggests that during 2015-16 the maximum share of area under rubber plantation to the total area under rubber plantation of the state is reported in West Tripura district with 23.71 percent of area under rubber plantation followed by Sepahijala, South Tripura, Gomati, North Tripura, Dhalai, Khowai and Unokoti districts. In the new administrative units, it is again West Tripura district which has the highest area under rubber plantation and is closely followed by Sepahijala and South Tripura districts.

To assess the concentration of area under rubber plantation in all the eight districts of the state for the year 2015-16, CI score has been computed. The CI score ranges from 2.64 to 0.25. Consequently, the concentration of area under rubber plantation for the eight districts has also been categorized into high concentration, moderate concentration, low concentration and very low concentration categories.

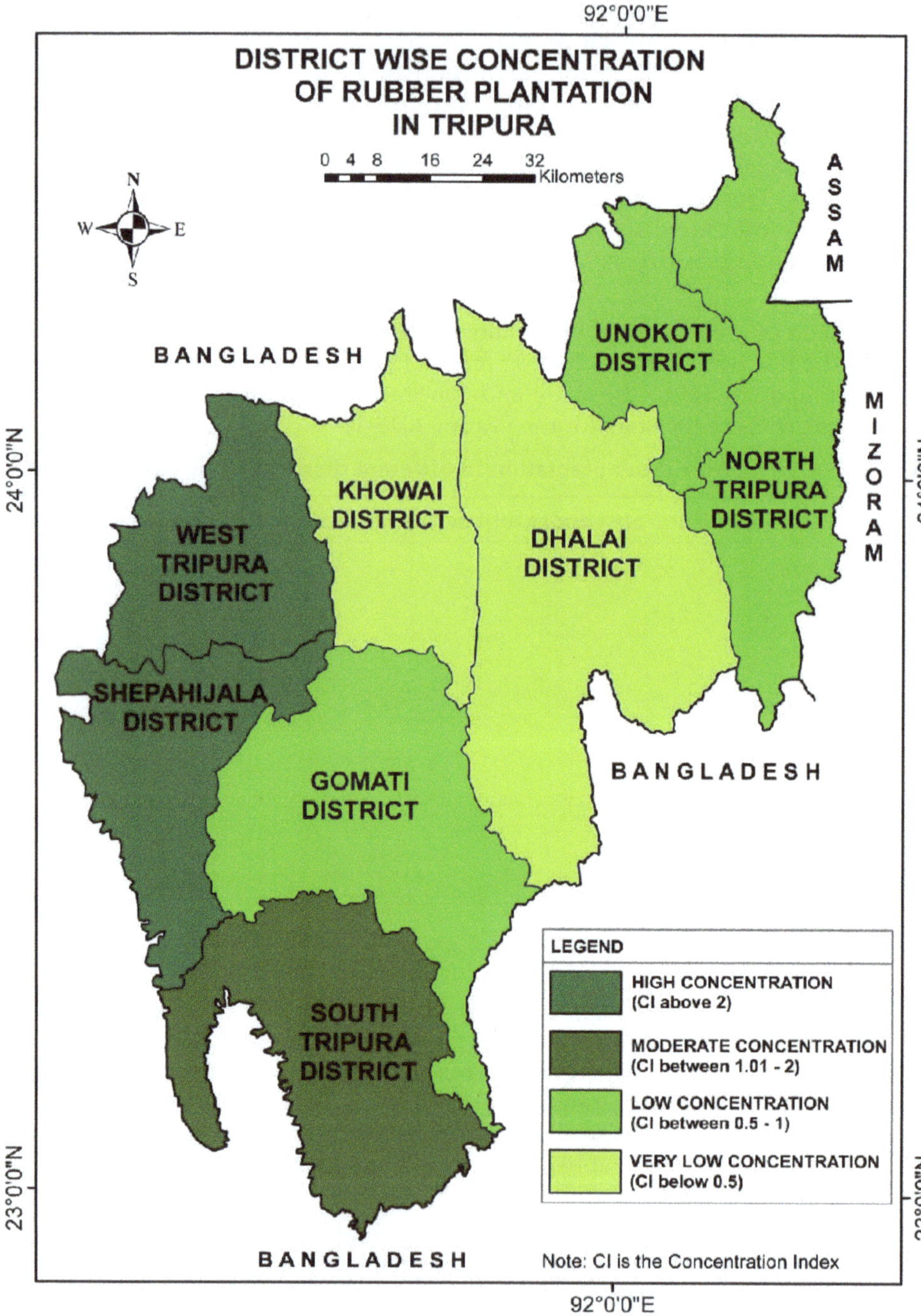

Fig. 3.12: District wise concentration of area under rubber plantation in Tripura.

It is observed from figure 3.12 that West Tripura and Sepahijala districts have the highest concentration of area under rubber plantation in the state with a CI score of 2.64 and 2.27 respectively. South Tripura district has moderate concentration of area under rubber plantation with a CI score of 1.42 and rank third after West Tripura and Sepahijala district. North Tripura, Gomati and Unokoti districts have low concentration of area under rubber plantation with a CI score of 0.70, 0.69, and 0.63 respectively. Khowai and Dhalai districts have very low concentration of area under rubber plantation with a CI score of 0.42 and 0.25 respectively.

1.6 Spatial distribution of rubber plantation in Tripura

RS and GIS technique has been used to map and analyze the spatial distribution of rubber plantation in Tripura using satellite data i.e. Landsat data of 2001 and 2015 as mentioned in methodology section of chapter I.

In order to examine the accuracy assessmeny of the image classification of rubber plantation for the year 2001 and 2015, the accuracy assessment formula of Raj et al. (2012) as mentioned in chapter I has been applied. Applying the formula, of the 50 no. of samples or references random point taken for classifying rubber plantation about 47 random points are correctly interpreted. This suggests that user accuracy of the classified image of rubber plantation for the year 2001 is about 94 percent accurate. On the other hand, in case for 2015, of the 50 no. of samples or references random points taken for rubber plantation classification a total of 48 random points is found to be correctly interpreted suggesting user accuracy of about 94.12 percent accuracy. These suggest that the classified image of rubber plantation from Landsat imageries 2001 and 2015 are highly accurate to the actual ground truth distribution of rubber plantation in the state.

The spatial distribution of rubber plantation in the state during March 2001 as derived from Landsat 5 TM image is depicted in figure 3.13. The figure indicates that rubber plantations are distributed mostly in the western and south parts of the state and towards the northern part of the state. Area under rubber plantation as of March 2001 estimated through RS is 18,024 ha which is less than the official data (23,197 ha) of the Rubber Board. In the satellite data young rubber plantation of about two years cannot be clearly marked as it often creates confusion with fallow/ scrub lands as the rubber trees canopy does not fully develop, consequently spectral signature becomes limited for the young plantation (Raj et al., 2012). Therefore, the rubber plantation area of young plantation (two years) has been added to the total rubber plantation area estimated through RS from the satellite data and the total area under rubber plantation of the state of Tripura for December 2011 matches with the official data of the Rubber Board. Here also the area of rubber plantation estimated through RS as of March 2001 as mentioned above is 18,024 ha and if added with the young rubber plantation area planted during 1998-99 and 1999-00 which is about 2,798 ha, the total area under rubber plantation as of March 2001 is estimated to be 20,822 ha. This almost matches with the official data available from the Rubber Board which is 23,197 ha as of March 2001 with slight variation (as the area under rubber plantation estimated through RS is slightly less than the actual data available with the Rubber Board).

The spatial distribution of rubber plantation in the state during March 2015 as derived from Landsat 8 OLI image (figure 3.14) also suggest that rubber plantation are concentrated mostly in the western and south parts of the state and towards the northern part of the state. For March 2015 as well the area under rubber plantation estimated through RS (57,328 ha) is added with the young rubber plantation area planted during 2012-13 and 2013-14 (4,908 ha) and the total area under rubber plantation as of March 2015 is about 62,236 ha suggesting slightly less than the official data (70,296 ha) of the Rubber Board. It is to be mentioned here that rubber plantation is usually planted from June to October/November consequently the total area estimated during March 2001 and 2015 is a correct reflection of the actual data of area under rubber plantation in the state.

The area under rubber plantation estimated through RS is less than the data available with the Rubber Board as those rubber plantations located in steep slopes are unaccounted due to limited/cannot be visualised from the satellite data consequently there is a high chance that the area under rubber plantation may be higher than the estimated data through remote sensing. On the other hand, the area under young rubber plantation can be more than the record provided by the Rubber Board which is submitted by the growers to the Rubber Board as there is a high chance that more rubber plantation exists than the submitted account of young rubber plantation to the Rubber Board by the growers (Raj et al., 2012).

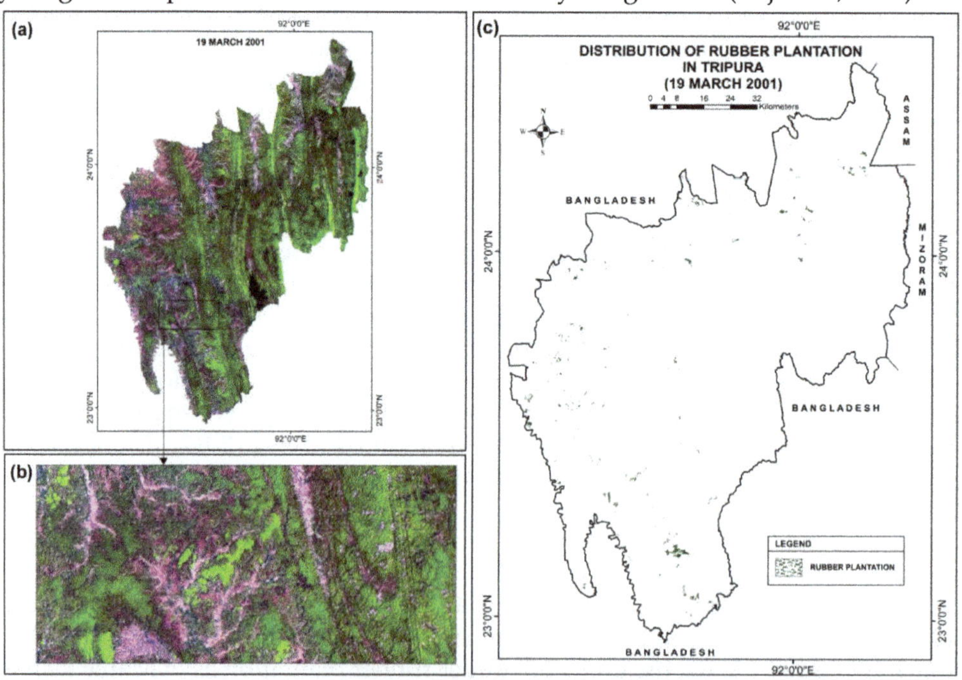

Fig. 3.13: (a) False colour composition map (R/G/B = Bands 5/4/3) of Landsat 5 TM image (2001), (b) the zoom-in image of a typical area show that rubber plantations are easily discernible as light green patches in the foliation stage in image and (c) show the spatial distribution of rubber plantation in the state.

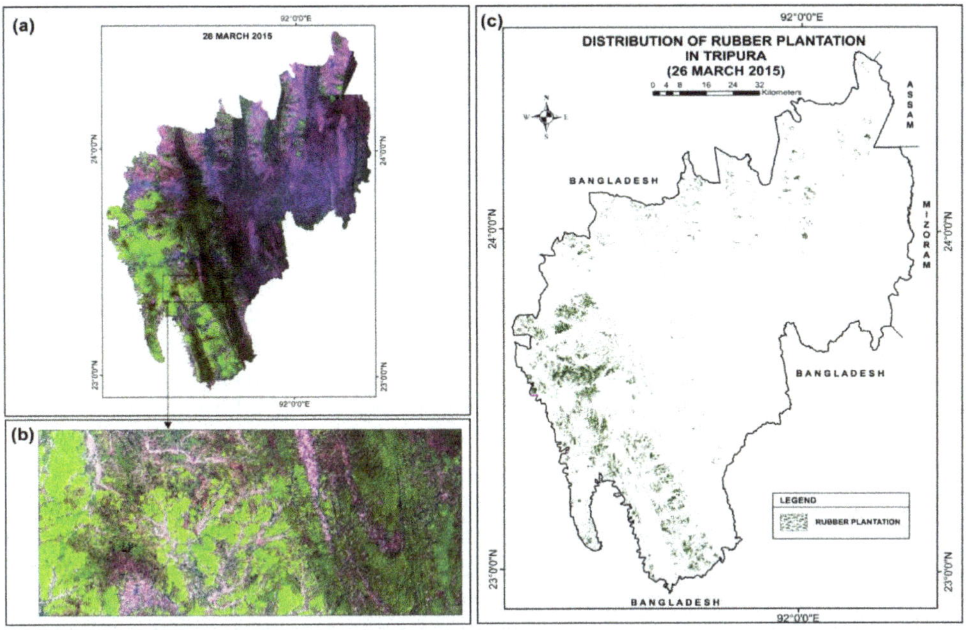

Fig. 3.14: (a) False colour composition map (R/G/B = Bands 6/5/4) of Landsat 8 OLI image (2015), (b) the zoom-in image of a typical area show that rubber plantations are easily discernible as light green patches in the foliation stage in image and (c) show the spatial distribution of rubber plantation in the state.

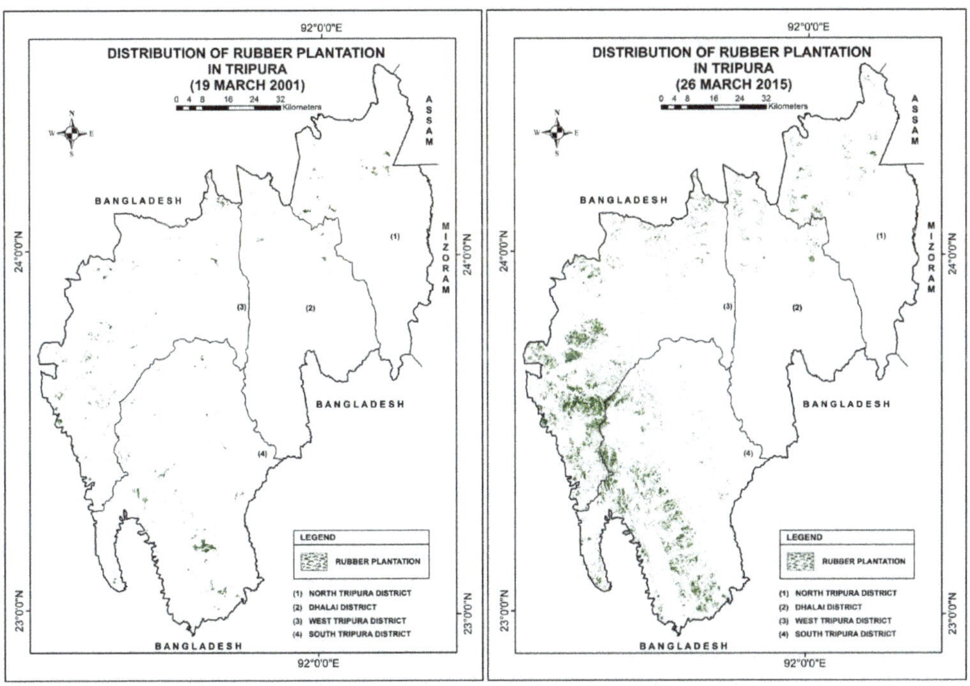

Fig. 3.15: Spatial distribution of rubber plantation in Tripura during 2001 and 2015.

It is observed from figure 3.15 that the spatial distribution of area under rubber plantation in the state of Tripura from 2001 to 2015 has increased in a similar pattern. The increase in the spatial distribution of area under rubber plantation from 2001 to 2015 are mainly in the West Tripura and South Tripura districts of the state and also towards the northern parts of the Dhalai and North Tripura districts of the state.

The agro-climatic conditions of the state are favourable for cultivation of rubber trees (Bhattacharya, 1992, cited in Maithani, 2005), however, cultivation of rubber trees is limited to certain areas where suitable locations exist. Jeschke and Strayer, 2008, cited in Ray, Behera and Jacob (2016) points out that the suitability of climatic conditions can't be the only deciding factor for cultivation of rubber trees but other factors like favourable condition of soil and topography are also considered necessary. Xiaona, Zhiming, Luguang, Peng, Chenhua, Yanzhao and Zhen (2013) highlights that the growing conditions and distribution of rubber plantation are influenced by a number of factors such as climate, topography, soil, hydrology, and socio-economic conditions. Ray, Behera and Jacob (2014) also highlights that climatically suitable areas for rubber trees are qualified on the basis of suitability of soil, physiography and socio-economic characteristics. Consequently, besides the agro-climatic conditions that has favoured the expansion of area under rubber plantation in the state, the influence of soil and physiography are also taken into account. Therefore the rubber plantation map has been overlayed on the land suitable for rubber plantation map and physiography map of the state of Tripura which is depicted in figure 3.16 and 3.17 respectively.

As depicted in figure 3.16 area under rubber plantations in the state are distributed mostly on the land moderately suitable for rubber plantation. As noticed most of the rubber plantations are distribution on the land that is considered 50 percent, 5 percent and 65 percent suitable for rubber plantation in the state. It is also noticed that rubber plantation has been established on the land considered 100 percent suitable for rubber plantation under the category of the marginally suitable land. The distribution of rubber plantation in relation to physiography as depicted in figure 3.17 indicates that most of the rubber plantation distributed has been established on the undulating plains, inter-hill valleys and rolling uplands as well as on flat topped topographic features suggesting that these are the lands considered suitable for rubber plantation in the state. It is noticed that rubber plantations are almost absent on the high relief topographic feature and also in most parts of the medium relief and low relief topographic features although rubber plantation has been established in few pockets on the low and medium relief topography as clearly shown in figure 3.17. Bhattacharyya, et al., (1998) also highlights that the moderately suitable areas for rubber plantation in the state of Tripura are mostly characterized by undulating plains and upland topography and also suggested that topography with less than 15 percent slope is considered suitable for rubber plantation. They also mention that in steep slope greater than 30-50 percent (although severe limitation for rubber plantation) rubber plantation can be established however with proper conservation measures.

Raj et al. (2012) has also indicates that the undulating terrain comprising of low to medium/gentle slope of the state are considered suitable for rubber plantation and also remarks that rubber plantation which has been setup in steep slopes have proper conservation measure like contour bunds management. It is in this context that some of the rubber plantations are noticed in some parts of the medium and low relief topographic features. This suggests that besides agro-climatic conditions, physiography also plays a dominant role in the distribution of rubber plantation in the state of Tripura.

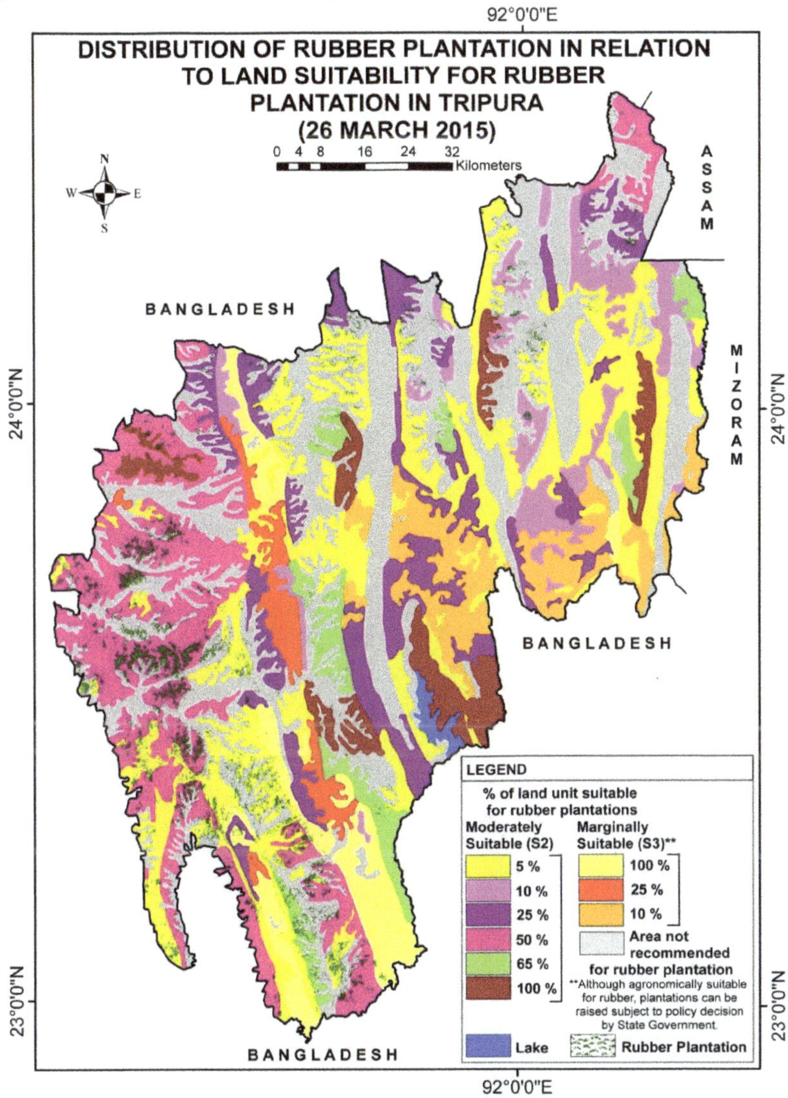

Fig. 3.16: Spatial distribution of rubber plantation in Tripura in relation to land suitability for rubber plantation (2015).

Source: Bhattacharyya, T., Sehgal, J and Sarkar, D., 1996, cited in Panagos, P., Jones, A., Bosco, C and Senthil Kumar, P.S. (2011).

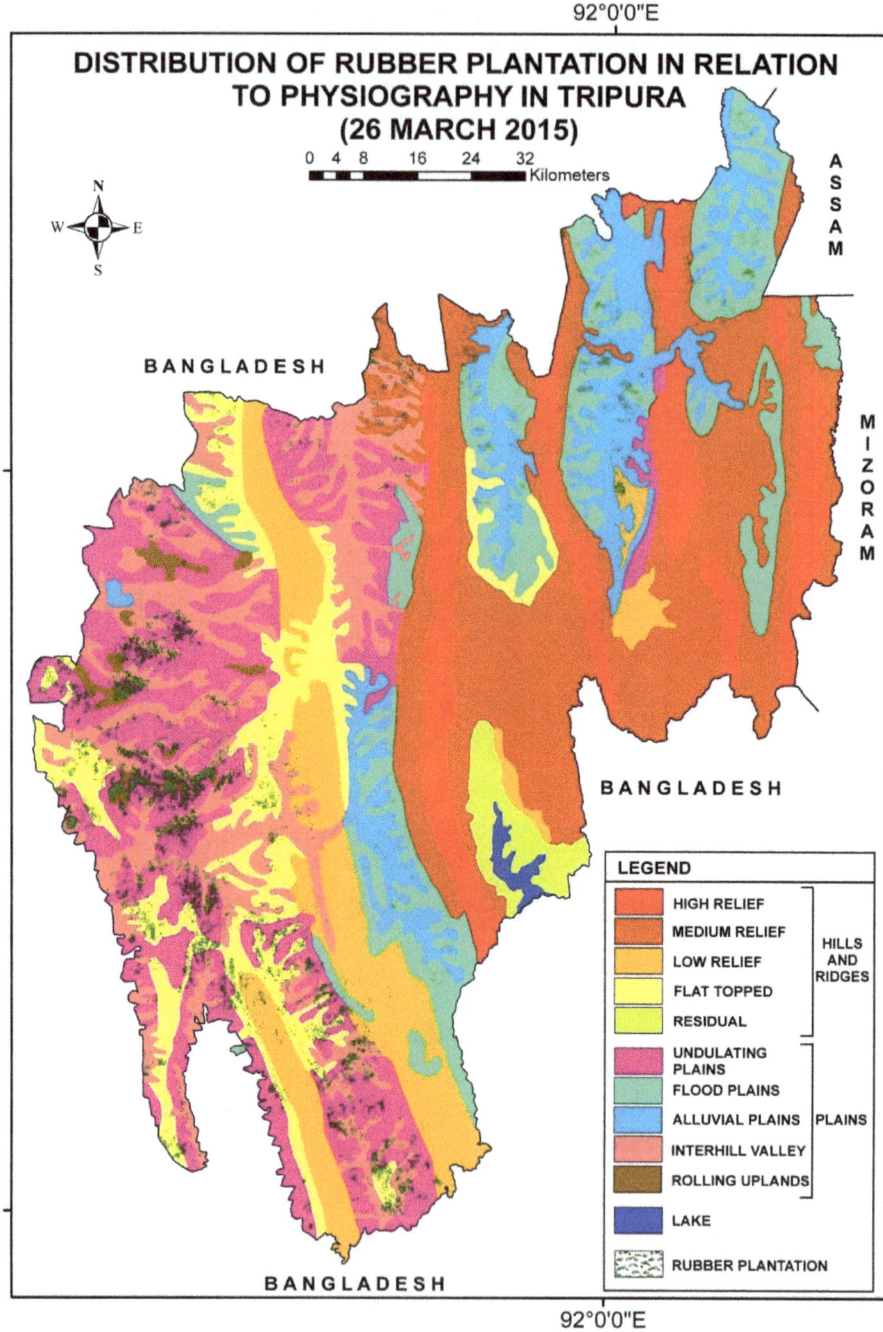

Fig. 3.17: Spatial distribution of rubber plantation in Tripura in relation to physiography (2015).

Source: Bhattacharyya, T., Sehgal, J and Sarkar, D., 1996, cited in Panagos, P., Jones, A., Bosco, C and Senthil Kumar, P.S. (2011).

As mentioned most of the rubber plantation has been established in the West Tripura and South Tripura districts of the state and the main reason for this are availability of suitable land for rubber plantation as in these districts the topography consists of undulating plains, inter-hill valleys, rolling uplands and flat topped features ideal for rubber plantation (figure 3.16 and 3.17). On the other hand, although rubber plantation in the northern parts of the Dhalai and North Tripura districts of the state has been established on the land moderately suitability for rubber plantation consisting of high and medium relief topography as well as plain areas. However, the topographic condition for rubber plantation in the northern parts of the state is restricted in comparison to western and southern parts of the state. Bhattacharyya et al. (1998) states that rubber plantations are restricted in high relief with steep slope and also in plain land as this land are prone to water stagnation harmful for rubber plantation.

Figure 3.16 suggest that rubber plantation is sparse of absent in the central and eastern parts of the state although this part of the state has land moderately suitable for rubber plantation. The main reason for this are hills and ridges consisting of high and medium relief (figure 3.17) covered with forest areas as rubber plantations are not encouraged on the steep slope and forested areas by the government.

1.7 Conclusion

Area under rubber plantation has been increasing continuously and spreading throughout the districts of the state. Area under rubber plantation in the state has been increasing rapidly but at present in a declining and fluctuating growth rate. Rubber plantations are highly concentrated in West Tripura district followed by South Tripura, North Tripura and Dhalai districts. There has been a similar spatial pattern in the distribution of rubber plantation in the state. Beside the favourable agro-climatic conditions influencing the expansion of rubber plantation in the state, physiography also has a dominant role in the distribution of rubber plantation in the state of Tripura. The success of rubber plantation in the state could not have been achieved without effective government policies and the acceptance of rubber plantation by the marginalized section of peoples inhabiting the rural areas. Consequently, rehabilitation through rubber plantation is reaching the rural areas of the state. It can be concluded that the expansion of area under rubber plantation in Tripura is a success due to the introduction of cash subsidy by the Rubber Board and the economic profitability. The rehabilitation programme of the *jhumias* by different state run agencies, increase in price of NR is also contributing factors for the expansion of area under rubber plantation in the state. Chattopadhyay (2015) notes that 'Globalisation, opening up of economy, high market price of rubber and emphasis on cash crop cultivation all together have contributed to the spatial spread' of rubber plantation in the state of Kerala. Similar factors may also have influence the expansion of area under rubber plantation throughout the traditional rubber growing and non-traditional rubber growing region of India including the state of Tripura. Subsequently it becomes important to assess the role of different agencies in popularising rubber plantation in the state of Tripura. The next chapter highlights on the different agencies, their objectives, and the role of such agencies in popularising rubber plantation in the state of Tripura.

References

Polhamus, L.G. 1962. *Rubber: Botany, Production, and Utilization*, Leonard Hill [Book] Limited, London, Interscience Publishers, INC., New York, pp. 1-30, 347-392.

Priyadarshan, P.M. 2011. *Biology of Hevea Rubber*, CAB International Publisher, pp.1-16.

Ibid, pp.1-16.

Polhamus, L.G. 1962. *Rubber: Botany, Production, and Utilization*, Leonard Hill [Book] Limited, London, Interscience Publishers, INC., New York, pp. 1-30, 347-392.

Priyadarshan, P.M. 2011. *Biology of Hevea Rubber*, CAB International Publisher, pp.1-16.

Polhamus, L.G. 1962. *Rubber: Botany, Production, and Utilization*, Leonard Hill [Book] Limited, London, Interscience Publishers, INC., New York, pp. 1-30, 347-392.

Priyadarshan, P.M. 2011. *Biology of Hevea Rubber*, CAB International Publisher, pp.1-16.

Ochse, J.J., Soule, J., Dijkman, M.J. and Wehlburg, C. 1961. *Tropical and Subtropical Agriculture, Volume II*, The Macmillan Company, Collier- Macmillan Ltd., London, pp. 923-996.

Ibid, pp. 923-996.

Kochhar, S.L. 1981. *Economic Botany in the Tropics*, Macmillan India Limited, Delhi, Bombay, Calcutta, Madras, pp. 385 – 395.

Priyadarshan, P.M. 2011. *Biology of Hevea Rubber*, CAB International Publisher, pp.1-16.

Ochse, J.J., Soule, J., Dijkman, M.J. and Wehlburg, C. 1961. *Tropical and Subtropical Agriculture, Volume II*, The Macmillan Company, Collier- Macmillan Ltd., London, pp. 923-996.

Cobley, L.S. 1976. *An Introduction to the Botany of Tropical Crops*, Longman Group Limited London, pp. 336-344.

Kochhar, S.L. 1981. *Economic Botany in the Tropics*, Macmillan India Limited, Delhi, Bombay, Calcutta, Madras, pp. 385-395.

Ochse, J.J., Soule, J., Dijkman, M.J. and Wehlburg, C. 1961. *Tropical and Subtropical Agriculture, Volume II*, The Macmillan Company, Collier- Macmillan Ltd., London, pp. 923-996.

Ibid, pp. 923-996.

Kochhar, S.L. 1981. *Economic Botany in the Tropics*, Macmillan India Limited, Delhi, Bombay, Calcutta, Madras, pp. 385-395.

Ochse, J.J., Soule, J., Dijkman, M.J. and Wehlburg, C. 1961. *Tropical and Subtropical Agriculture, Volume II*, The Macmillan Company, Collier- Macmillan Ltd., London, pp. 923-996.

Ibid, pp. 923-996.

Kochhar, S.L. 1981. *Economic Botany in the Tropics*, Macmillan India Limited, Delhi, Bombay, Calcutta, Madras, pp. 385 – 395.

Polhamus, L.G. 1962. *Rubber: Botany, Production, and Utilization*, Leonard Hill [Book] Limited, London, Interscience Publishers, INC., New York, pp. 1-30, 347-392.

Peter, K.V. 2002. *Plantation Crops*, National Book Trust, New Delhi, pp. 247-277. Mors, W.B. and Rizzini, C.T. 1966. *Useful Plants of Brazil*, Holden-Day Inc., San

Francisco, London, Amsterdam, pp. 1-7.

Kochhar, S.L. 1981. *Economic Botany in the Tropics*, Macmillan India Limited, Delhi, Bombay, Calcutta, Madras, pp. 385-395.

Opeke, L.K. 1982. *Tropical Tree Crops*, John Wiley and Sons Ltd., Chichester, New York, Brisbane, Toronto, Singapore, pp. 216-240.

Peter, K.V. 2002. *Plantation Crops*, National Book Trust, New Delhi, pp. 247-277. Ranganath, B.K., PRADEEP, N., Manjula, V.B., Gowda, B., Rajanna, M.D.,

Shettigar, D., Rao, P.P.N. 2004. Detection of diseased rubber plantations using satellite remote sensing, *Journal of the Indian Society of Remote Sensing, Vol.32, No.1*, pp. 49-58.

Gleason, C. 2005. *The Biography of Rubber*, Crabtree Publishing Company, pp. 1-32. Raj, U., Hebbar, R., Ravishankar, H.M., Jacob, J., Ray, D., Meti, S., Shebin, S.M., and Pradeep, B. 2012. Geospatial Technology for Acreage Estimation of Natural Rubber and Identification of Potential Areas for its Cultivation in Tripura, *Project Report, Issue no. 1*, National Remote Sensing Centre, ISRO,

Govt. of India and Rubber Research Institute of India, Rubber Board, Min. of Commerce & Industry, Govt. of India, pp. 1-45.

Ochse, J.J., Soule, J., Dijkman, M.J. and Wehlburg, C. 1961. *Tropical and Subtropical Agriculture, Volume II*, The Macmillan Company, Collier- Macmillan Ltd., London, pp. 923-996.

Polhamus, L.G. 1962. *Rubber: Botany, Production, and Utilization*, Leonard Hill [Book] Limited, London, Interscience Publishers, INC., New York, pp. 1-30, 347-392.

Mors, W.B. and Rizzini, C.T. 1966. *Useful Plants of Brazil*, Holden-Day Inc., San Francisco, London, Amsterdam, pp. 1-7.

Peter, K.V. 2002. *Plantation Crops*, National Book Trust, New Delhi, pp. 247-277. Roy, P. 2005. *Economic Geography– A Study of Resources*, New Central Book

Agency (P) Ltd., Kolkata ,pp. 205-209, 492.

Priyadarshan,P.M. 2011. *Biology of Hevea Rubber*, CAB International Publisher, pp.1-16.

Peter, K.V. 2002. *Plantation Crops*, National Book Trust, New Delhi, pp. 247-277.

Borthakur, D.N. 1992. *Agriculture of the North eastern region: with special references to hill Agriculture*, BeeceePrakashan, Guwahati, pp. 59-71.

Raj, U., Hebbar, R., Ravishankar, H.M., Jacob, J., Ray, D., Meti, S., Shebin, S.M. and Pradeep, B. 2012. Geospatial Technology for Acreage Estimation of Natural Rubber and Identification of Potential Areas for its Cultivation in Tripura, *Project Report, Issue no. 1,* National Remote Sensing Centre, ISRO, Govt. of India and Rubber Research Institute of India, Rubber Board, Min. of Commerce & Industry, Govt. of India, pp. 1-45.

Rubber Board. *The Rubber Tree.* Retrieved from http://rubberboard.org.in/rubbercultivation.asp. Accessed on April 2015.

Rubber Board. *The Rubber Tree.* Retrieved from http://rubberboard.org.in/rubbercultivation.asp. Accessed on April 2015.

George K.. T., Haridasan, V. and Sreekumar, B. 1988. Role of Government and Structural Changes in Rubber Plantation Industry, *Economic and Political Weekly,Vol. 23, No. 48,* pp. M158-M166.

Kochhar, S.L. 1981. *Economic Botany in the Tropics,* Macmillan India Limited, Delhi, Bombay, Calcutta, Madras, pp. 385-395.

Priyadarshan,P.M. 2011. *Biology of Hevea Rubber,* CAB International Publisher, pp.1-16.

Kochhar, S.L. 1981. *Economic Botany in the Tropics,* Macmillan India Limited, Delhi, Bombay, Calcutta, Madras, pp. 385-395.

Priyadarshan,P.M. 2011. *Biology of Hevea Rubber,* CAB International Publisher, pp.1-16.

Ibid, pp.1-16.

Tiwari, R.C. 2008. *Geography of India,* Prayag Pustak Bhawan, Allahabad, pp. 344-346.

Roy, P. 2005. *Economic Geography– A Study of Resources,* New Central Book Agency (P) Ltd., Kolkata, pp. 205-209, 492.

George K. T., Haridasan, V. and Sreekumar, B. 1988. Role of Government and Structural Changes in Rubber Plantation Industry, *Economic and Political Weekly,Vol. 23, No. 48,* pp. M158-M166.

Polhamus, L.G. 1962. *Rubber: Botany, Production, and Utilization,* Leonard Hill [Book] Limited, London, Interscience Publishers, INC., New York, pp. 1-30, 347-392.

Sinha, A.K. 2010. Prospect of rubber plantation in NE region with special reference to the state of Tripura, *The Tripura Foundation.* Retrieved from http://thetripurafoundation.org/art-rubber-prospect-tripura-and-north-east.

Accessed on April 2015.

Polhamus, L.G. 1962. *Rubber: Botany, Production, and Utilization,* Leonard Hill [Book] Limited, London, Interscience Publishers, INC., New York, pp. 1-30, 347-392.

Raghavan, K.K. 2009. Rubber for tribal welfare in Tripura, in G.K. Bera, B.K. Medhi, R.P. Athparia, and K. Jose (ed.) *Tribal Development in Tripura,* EBH Publisher, Guwahati, pp. 89-98.

Polhamus, L.G. 1962. *Rubber: Botany, Production, and Utilization,* Leonard Hill [Book] Limited, London, Interscience Publishers, INC., New York, pp. 1-30, 347-392.

George K.. T., Haridasan, V. and Sreekumar, B. 1988. Role of Government and Structural Changes in Rubber Plantation Industry, *Economic and Political Weekly,Vol. 23, No. 48,* pp. M158-M166.

Ibid, pp. M158-M166.

George K.. T., Haridasan, V. and Sreekumar, B. 1988. Role of Government and Structural Changes in Rubber Plantation Industry, *Economic and Political Weekly,Vol. 23, No. 48,* pp. M158-M166.

Rajasenan, D. 2010. Livelihood and employment of workers in rubber and spices plantations, *NRPPD Discussion Paper- 6,* pp. 1-68.

DESPD. 2016. *Economic Review of Tripura (2015-16), 17th issue,* Government of Tripura, Agartala, Tripura, pp. 3-151.

Ibid, pp. 3-151.

Vinod, K.K., Meenattor, J.R., Priyadarshan, P.M., Pothen, J., Chaudhuri, D., Krishnakumar, A.K., Sethuraj, M.R. and Potty, S.N. 1996. Early Performance of Some Clones of Hevea Brasiliensis in Tripura, *Indian Journal of Natural Rubber Research, 9(2),* pp. 123-129.

Bhowmik, I. 2006. A status on Rubber Plantations in Tripura, *Tripura Rubber Mission- Technical Bulletin,* Tripura, pp. 1-21.

Vinod, K.K., Meenattoor, J.R., Reddy, Y.A., Priyadarshan, P.M., Chaudhuri, D. 2010. Ontogenetic variations in flush development are indicative of low temperature tolerance in *Hevea brasiliensis* clones, *Annals of Forest Research, 53(2),* pp. 95-105.

Raj, U., Hebbar, R., Ravishankar, H.M., Jacob, J., Ray, D., Meti, S., Shebin, S.M. and Pradeep, B. 2012. Geospatial Technology for Acreage Estimation of Natural Rubber and Identification of Potential Areas for its Cultivation in Tripura, *Project Report, Issue no. 1,* National Remote Sensing Centre, ISRO, Govt. of India and Rubber Research Institute of India, Rubber Board, Min. of Commerce & Industry, Govt. of India, pp. 1-45.

Joseph, J., George K, T. and Dey, S.K. 2010. Report on the socio-economic impact of Natural Rubber cultivation under the block planting scheme in Tripura, *MPRA Paper No. 52370*: 1-39.

Sharma, G., Joseph, J., George K, T. and Dey, S.K. 2011. Impact of Mahatma Gandhi National Rural Employment Guarantee Act on Rubber Block Plantation Scheme in Tripura, *Agriculture Economic Research Review Vol. 24*: 525-530.

Borthakur, D.N. 1992. *Agriculture of the North eastern region: with special references to hill Agriculture,* BeeceePrakashan, Guwahati, pp. 59-71.

Bhowmik, I. 2006. A status on Rubber Plantations in Tripura, *Tripura Rubber Mission- Technical Bulletin,* Tripura, pp. 1-21.

Vinod, K.K., Meenattor, J.R., Priyadarshan, P.M., Pothen, J., Chaudhuri, D., Krishnakumar, A.K., Sethuraj, M.R. and Potty, S.N. 1996. Early Performance of Some Clones of Hevea Brasiliensis in Tripura, *Indian Journal of Natural Rubber Research, 9(2)*, pp. 123-129.

Borthakur, D.N. 1992. *Agriculture of the North eastern region: with special references to hill Agriculture*, BeeceePrakashan, Guwahati, pp. 59-71.

Viswanathan, P.K. and Bhowmik, I. 2014. Compatibility of Institutional Architecture for Rubber Plantation Development in North East India from the Comparative Perspective of Kerala, *NRPPD Discussion Paper No. 38*, Centre For Development Studies, Thiruvananthapuram, pp. 1-71.

Sarkar, A.N. 1994. *Integrated Horticulture Development in Eastern Himalayas*, M.D.Publication Pvt. Ltd., New Delhi, pp. 114-123.

Gupta, A.K. 2000. Shifting Cultivation and Conservation of Biological Diversity in Tripura, Northeast India, *Human Ecology, Vol. 28, No. 4*,pp. 605-629.

Centre for Development Studies. *Area under rubber and average yield per hectare.* Retrieved from http://cds.edu/research/ru/nrppd/database-on- plantation/ rubber/. Accessed on 23 January, 2016.

Vinod, K.K., Meenattor, J.R., Priyadarshan, P.M., Pothen, J., Chaudhuri, D., Krishnakumar, A.K., Sethuraj, M.R. and Potty, S.N. 1996. Early Performance of Some Clones of Hevea Brasiliensis in Tripura, *Indian Journal of Natural Rubber Research, 9(2)*, pp. 123-129.

Bhowmik, I. 2006. A status on Rubber Plantations in Tripura, *Tripura Rubber Mission- Technical Bulletin*, Tripura, pp. 1-21.

Viswanathan, P.K. and Bhowmik, I. 2014. Compatibility of Institutional Architecture for Rubber Plantation Development in North East India from the Comparative Perspective of Kerala, *NRPPD Discussion Paper No. 38*, Centre For Development Studies, Thiruvananthapuramy pp. 1-71.

Chakravarti, M. 1986. Forestry in Tripura: Its Past and Present, in M.D. Gupta, A.K. Gangopadhyay, T. Bhattacharya, M. Chakraborti (ed.) *Forestry Development in North-East India*, Omsons Publications, Guwahati and New Delhi, pp. 31- 41.

Bhowmik, I. 2006. A status on Rubber Plantations in Tripura, *Tripura Rubber Mission- Technical Bulletin*, Tripura, pp. 1-21.

Raj, U., Hebbar, R., Ravishankar, H.M., Jacob, J., Ray, D., Meti, S., Shebin, S.M. and Pradeep, B. 2012. Geospatial Technology for Acreage Estimation of Natural Rubber and Identification of Potential Areas for its Cultivation in Tripura, *Project Report, Issue no. 1*, National Remote Sensing Centre, ISRO, Govt. of India and Rubber Research Institute of India, Rubber Board, Min. of Commerce & Industry, Govt. of India, pp. 1-45.

Bhowmik, I. 2006. A status on Rubber Plantations in Tripura, *Tripura Rubber Mission- Technical Bulletin*, Tripura, pp. 1-21.

Viswanathan, P.K. and Bhowmik, I. 2014. Compatibility of Institutional Architecture for Rubber Plantation Development in North East India from the Comparative Perspective of Kerala, *NRPPD Discussion Paper No. 38,* Centre For Development Studies, Thiruvananthapuram, pp. 1-71.

Ibid, pp. 1-71.

Bhowmik, I. 2006. A status on Rubber Plantations in Tripura, *Tripura Rubber Mission- Technical Bulletin,* Tripura, pp. 1-21.

Viswanathan, P.K. and Bhowmik, I. 2014. Compatibility of Institutional Architecture for Rubber Plantation Development in North East India from the Comparative Perspective of Kerala, *NRPPD Discussion Paper No. 38,* Centre For Development Studies, Thiruvananthapuram, pp. 1-71.

Bhowmik, I. 2006. A status on Rubber Plantations in Tripura, *Tripura Rubber Mission- Technical Bulletin,* Tripura, pp. 1-21.

Joseph, J., George K, T. and Dey, S.K. 2010. Report on the socio-economic impact of Natural Rubber cultivation under the block planting scheme in Tripura, *MPRA Paper No. 52370,* pp. 1-39.

Viswanathan, P.K. and Bhowmik, I. 2014. Compatibility of Institutional Architecture for Rubber Plantation Development in North East India from the Comparative Perspective of Kerala, *NRPPD Discussion Paper No. 38,* Centre For Development Studies, Thiruvananthapuram, pp. 1-71.

Sarkar, M. 1986. Forestry Development in Tripura During the Five Year Plans, in M.D. Gupta, A.K. Gangopadhyay, T. Bhattacharya, M. Chakraborti (ed.) *Forestry Development in North-East India,* Omsons Publications, Guwahati and New Delhi, pp. 192-203.

Bhowmik, I. 2006. A status on Rubber Plantations in Tripura, *Tripura Rubber Mission- Technical Bulletin,* Tripura, pp. 1-21.

Raghavan, K.K. 2009. Rubber for tribal welfare in Tripura, in G.K. Bera, B.K. Medhi, R.P. Athparia, and K. Jose (ed.) *Tribal Development in Tripura,* EBH Publisher, Guwahati, pp. 89-98.

Choudhury, J. 2009. Efforts of Tribal Development in Tripura: An Assessment, in G.K. Bera, B.K. Medhi, R.P. Athparia, and K. Jose (ed.) *Tribal Development in Tripura,* EBH Publisher, Guwahati, pp. 121-135.

Viswanathan, P.K. and Bhowmik, I. 2014. Compatibility of Institutional Architecture for Rubber Plantation Development in North East India from the Comparative Perspective of Kerala, *NRPPD Discussion Paper No. 38,* Centre For Development Studies, Thiruvananthapuram, pp. 1-71.

Bhowmik, I. 2006. A status on Rubber Plantations in Tripura, *Tripura Rubber Mission- Technical Bulletin,* Tripura, pp. 1-21.

Raghavan, K.K. 2009. Rubber for tribal welfare in Tripura, in G.K. Bera, B.K. Medhi, R.P. Athparia, and K. Jose (ed.) *Tribal Development in Tripura,* EBH Publisher, Guwahati, pp. 89-98.

Bhattacharjee, S 2012. Multidisciplinary research can make the medicinal plant industry in Tripura viable- A conceptual study based on tows matrix, *International Journal of Multidisciplinary Research, Vol 2, Issue 5,* pp. 146- 154.

Sharma, G., George K., T. and Dey, S.K. 2014. Natural Rubber Sector in Tripura: Role of Institutional Innovations, *Agricultural Economics Research Review, Vol. 27*, pp 17-24.

Viswanathan, P.K. and Bhowmik, I. 2014. Compatibility of Institutional Architecture for Rubber Plantation Development in North East India from the Comparative Perspective of Kerala, *NRPPD Discussion Paper No. 38*, Centre For Development Studies, Thiruvananthapuram, pp. 1-71.

Rubber Board. 2016. *Role of Rubber Board in Growth of Rubber Plantation in NE*, The regional office, Agartala.

DESPD. 2016. *Economic Review of Tripura (2015-16), 17th issue*, Government of Tripura, Agartala, Tripura, pp. 3-151.

Bhowmik, I. 2006. A status on Rubber Plantations in Tripura, *Tripura Rubber Mission- Technical Bulletin*, Tripura, pp. 1-21.

Viswanathan, P.K. and Bhowmik, I. 2014. Compatibility of Institutional Architecture for Rubber Plantation Development in North East India from the Comparative Perspective of Kerala, *NRPPD Discussion Paper No. 38*, Centre For Development Studies, Thiruvananthapuram, pp. 1-71.

DESPD. 2008. *Economic Review of Tripura (2007-08), 9th issue*, Government of Tripura, Agartala, Tripura, pp. 3-153.

Devi, A.I. 2010. *Amazing North-East Tripura*, Vij Books India Pvt. Ltd. New Delhi, pp. 75-77.

Sharma, G., Joseph, J., George K, T. and Dey, S.K. 2011. Impact of Mahatma Gandhi National Rural Employment Guarantee Act on Rubber Block Plantation Scheme in Tripura, *Agriculture Economic Research Review Vol. 24*: 525-530.

Chakravarti, M. 1986. Forestry in Tripura: Its Past and Present, in M.D. Gupta, A.K. Gangopadhyay, T. Bhattacharya, M. Chakraborti (ed.) *Forestry Development in North-East India*, Omsons Publications, Guwahati and New Delhi, pp. 31-41.

Bhowmik, I. 2006. A status on Rubber Plantations in Tripura, *Tripura Rubber Mission- Technical Bulletin*, Tripura, pp. 1-21.

Maithani, B.P., 2005. *Shifting Cultivation in North-East India: Policy issues and options*, Mittal Publication, New Delhi. pp. 30-50.

Raghavan, K.K. 2009. Rubber for tribal welfare in Tripura, in G.K. Bera, B.K. Medhi, R.P. Athparia, and K. Jose (ed.) *Tribal Development in Tripura*, EBH Publisher, Guwahati, pp. 89-98.

Raj, U., Hebbar, R., Ravishankar, H.M., Jacob, J., Ray, D., Meti, S., Shebin, S.M. and Pradeep, B. 2012. Geospatial Technology for Acreage Estimation of Natural Rubber and Identification of Potential Areas for its Cultivation in Tripura, *Project Report, Issue no. 1*, National Remote Sensing Centre, ISRO, Govt. of India and Rubber Research Institute of India, Rubber Board, Min. of Commerce & Industry, Govt. of India, pp. 1-45.

DESPD. 2015. *Economic Review of Tripura (2014-15), 16th issue,* Government of Tripura, Agartala, Tripura, pp. 3-157.

Bhowmik, I. and Viswanathan, P.K. 2015. Emerging Labour Relations in the Small Rubber Plantations of Tripura, *NRPPD, Discussion Paper No. 47*, Centre For Development Studies, Thiruvananthapuram, pp. 1-33.

Centre for Development Studies. *Area under rubber and average yield per hectare.* Retrieved from http://cds.edu/research/ru/nrppd/database-on- plantation/ rubber/. Accessed on 23 January 2016.

Rubber Board. 2016. *Role of Rubber Board in Growth of Rubber Plantation in NE,* The regional office, Agartala.

Das, R. 2014. Rubber Plantation: Tripura Scenario, in K. Shib (ed.) *Rubber in Tripura (1963-2013)- Golden Jubilee Celebration, Souvenir,* Bina Printers, Arundhatinagar, Agartala, pp. 25-33.
Rubber Board. 2016. *Role of Rubber Board in Growth of Rubber Plantation in NE,* The regional office, Agartala.

Centre for Development Studies. *Area under rubber and average yield per hectare.* Retrieved from http://cds.edu/research/ru/nrppd/database-on- plantation/ rubber/. Accessed on 23 January 2016.
Viswanathan, P.K. and Bhowmik, I.,2014. Compatibility of Institutional Architecture for Rubber Plantation Development in North East India from the Comparative Perspective of Kerala, *NRPPD Discussion Paper No. 38,* Centre For Development Studies, Thiruvananthapuram, pp. 1-71.

Ibid, pp. 1-71.

Ibid, pp. 1-71.

Mohanakumar, S. 2013. Expansion of Natural Rubber Cultivation in Tripura: Impact on Landholding, Employment and Income, *NLI Research Studies Series No. 107,* pp. 1-46.

Mohanakumar, S. 2014. Political Economy of Natural Rubber Cultivation in Tripura, *IDSJ Working Paper 172,* Institute of Development Studies, Jaipur (INDIA), pp. 1-16.

Bhowmik, I. 2006. A status on Rubber Plantations in Tripura, *Tripura Rubber Mission- Technical Bulletin,* Tripura, pp. 1-21.

Joseph, J., George K, T. and Dey, S.K. 2010. Report on the socio-economic impact of Natural Rubber cultivation under the block planting scheme in Tripura, *MPRA Paper No. 52370,* pp. 1-39.

Viswanathan, P.K. and Bhowmik, I. 2014. Compatibility of Institutional Architecture for Rubber Plantation Development in North East India from the Comparative Perspective of Kerala, *NRPPD Discussion Paper No. 38,* Centre For Development Studies, Thiruvananthapuram, pp. 1-71.
Centre for Development Studies. *Area under rubber and average yield per hectare.* Retrieved from http://cds.edu/research/ru/nrppd/database-on- plantation/ rubber/. Accessed on 23 January 2016.

Rubber Board. 2016. *Role of Rubber Board in Growth of Rubber Plantation in NE,* The regional office, Agartala.

Mohanakumar, S. 2013. Expansion of Natural Rubber Cultivation in Tripura: Impact on Landholding, Employment and Income, *NLI Research Studies Series No. 107,* pp. 1-46.

Ibid, pp. 1-46.

Ibid, pp. 1-46.

Ibid, pp. 1-46.

Centre for Development Studies. *Area under rubber and average yield per hectare.* Retrieved from http://cds.edu/research/ru/nrppd/database-on- plantation/ rubber/. Accessed on 23 January 2016.

Rubber Board. 2016. *Role of Rubber Board in Growth of Rubber Plantation in NE,* The regional office, Agartala.

DESPD. 2009. *Economic Review of Tripura (2008-09), 10th issue,* Government of Tripura, Agartala, Tripura, pp. 3-142.

DESPD. 2010. *Economic Review of Tripura (2009-10), 11th issue,* Government of Tripura, Agartala, Tripura, pp. 3-131.

DESPD. 2011. *Economic Review of Tripura (2010-11), 12th issue,* Government of Tripura, Agartala, Tripura, pp. 3-151.

DESPD. 2012. *Economic Review of Tripura (2011-12), 13th issue,* Government of Tripura, Agartala, Tripura, pp. 3-170.

DESPD. 2013. *Economic Review of Tripura (2012-13), 14th issue,* Government of Tripura, Agartala, Tripura, pp. 3-152.

DESPD. 2014. *Economic Review of Tripura (2013-14), 15th issue,* Government of Tripura, Agartala, Tripura, pp. 3-157.

DESPD. 2015. *Economic Review of Tripura (2014-15), 16th issue,* Government of Tripura, Agartala, Tripura, pp. 3-157.

DESPD. 2016. *Economic Review of Tripura (2015-16), 17th issue,* Government of Tripura, Agartala, Tripura, pp. 3-151.

Rubber Board. 2016. *Role of Rubber Board in Growth of Rubber Plantation in NE,* The regional office, Agartala.

Bhowmik, I. 2006. A status on Rubber Plantations in Tripura, *Tripura Rubber Mission- Technical Bulletin,* Tripura, pp. 1-21.

Viswanathan, P.K. and Bhowmik, I. 2014. Compatibility of Institutional Architecture for Rubber Plantation Development in North East India from the Comparative Perspective of Kerala, *NRPPD Discussion Paper No. 38,* Centre For Development Studies, Thiruvananthapuram, pp. 1-71.

Census of India. 2011. *2011 census data-Population Enumeration Data,* Office of the Registrar General and Census Commissioner, India, New Delhi. Retrieved from http://www.censusindia.gov.in/2011census/population_enumeration.html. Accessed on 4 February 2016.

DESPD. 2009. *Economic Review of Tripura (2008-09), 10th issue,* Government of Tripura, Agartala, Tripura, pp. 3-142.

DESPD. 2010. *Economic Review of Tripura (2009-10), 11th issue,* Government of Tripura, Agartala, Tripura, pp. 3-131.

DESPD. 2011. *Economic Review of Tripura (2010-11), 12th issue,* Government of Tripura, Agartala, Tripura, pp. 3-151.

DESPD. 2012. *Economic Review of Tripura (2011-12), 13th issue,* Government of Tripura, Agartala, Tripura, pp. 3-170.

DESPD. 2013. *Economic Review of Tripura (2012-13), 14th issue,* Government of Tripura, Agartala, Tripura, pp. 3-152.

DESPD. 2014. *Economic Review of Tripura (2013-14), 15th issue,* Government of Tripura, Agartala, Tripura, pp. 3-157.

DESPD. 2015. *Economic Review of Tripura (2014-15), 16th issue,* Government of Tripura, Agartala, Tripura, pp. 3-157.

DESPD. 2016. *Economic Review of Tripura (2015-16), 17th issue,* Government of Tripura, Agartala, Tripura, pp. 3-151.

Rubber Board. 2016. *Role of Rubber Board in Growth of Rubber Plantation in NE,* The regional office, Agartala.

Mohanakumar, S. 2014. Political Economy of Natural Rubber Cultivation in Tripura, *IDSJ Working Paper 172,* Institute of Development Studies, Jaipur (INDIA), pp. 1-16.

TRPCL. 2012. *Brief Note on Tripura Rehabilitation Plantation Corporation Ltd. Diary 2012,* Agartala, Tripura, pp. 1-12.

TRPCL. 2016. *Documents on Tripura Rehabilitation Plantation Corporation Limited,* Agartala, Tripura.

Raj, U., Hebbar, R., Ravishankar, H.M., Jacob, J., Ray, D., Meti, S., Shebin, S.M. and Pradeep, B. 2012. Geospatial Technology for Acreage Estimation of Natural Rubber and Identification of Potential Areas for its Cultivation in Tripura, *Project Report, Issue no. 1,* National Remote Sensing Centre, ISRO, Govt. of India and Rubber Research Institute of India, Rubber Board, Min. of Commerce & Industry, Govt. of India, pp. 1-45.

Mohanakumar, S. 2013. Expansion of Natural Rubber Cultivation in Tripura: Impact on Landholding, Employment and Income, *NLI Research Studies Series No. 107,* pp. 1-46.

DESPD. 2012. *Economic Review of Tripura (2011-12), 13th issue,* Government of Tripura, Agartala, Tripura, pp. 3-170.

Rubber Board. 2016. *Role of Rubber Board in Growth of Rubber Plantation in NE,* The regional office, Agartala.

Raj, U., Hebbar, R., Ravishankar, H.M., Jacob, J., Ray, D., Meti, S., Shebin, S.M. and Pradeep, B. 2012. Geospatial Technology for Acreage Estimation of Natural Rubber and Identification of Potential Areas for its Cultivation in Tripura, *Project Report, Issue no. 1,* National Remote Sensing Centre, ISRO,

Govt. of India and Rubber Research Institute of India, Rubber Board, Min. of Commerce & Industry, Govt. of India, pp. 1-45.

Ibid, pp. 1-45.

Ibid, pp. 1-45.

Maithani, B.P. 2005. *Shifting Cultivation in North-East India: Policy issues and options*, Mittal Publication, New Delhi, pp. 30-50.

Ray, D., Beehera, M.D.and Jacob, J. 2016. Predicting the distribution of rubber trees (Hevea brasiliensis) through ecological niche modeling with climate, soil, topography and socioeconomic factors, *Ecol Res, 31*, pp. 75-91.

Xiaona, L., Zhiming, F., Luguang, J., Peng, L., Chenhua, L., Yanzhao, Y. and Zhen, Y. 2013. Rubber plantation and its relationship with topographical factors in the border region of China, Laos and Myanmar, *Journal of Geographical Sciences, Vol 23 (6)*, pp. 1019-1040.

Ray, D., Behera, M.D. and Jacob, J. 2014. Indian Brahmaputra valley offers significant potential for cultivation of rubber trees under change climate, *Current Science, Vol. 107, No. 3*, pp. 461-469.

Bhattacharyya, T., Sarkar, D., Gangopadhyay, S.K., Dubey, P.N., Baruah, U., Chamuah, G.S., Mukhopadhyay, S., Nayak, D.C., Maji, A.K., Saxena, R.K., Barthwal, A.K., Krishna, N.D.R., Mandal, C. and Sahgal, J. 1998. Soils of Tripura.II- Suitability for Rubber, *Agropedology, 8*, pp. 55-60.

Raj, U., Hebbar, R., Ravishankar, H.M., Jacob, J., Ray, D., Meti, S., Shebin, S.M. and Pradeep, B. 2012. Geospatial Technology for Acreage Estimation of Natural Rubber and Identification of Potential Areas for its Cultivation in Tripura, *Project Report, Issue no. 1*, National Remote Sensing Centre, ISRO, Govt. of India and Rubber Research Institute of India, Rubber Board, Min. of Commerce & Industry, Govt. of India, pp. 1-45.

Panagos, P., Jones, A., Bosco, C. and Senthil Kumar, P.S. 2011, European digital archive on soil maps (EuDASM): preserving important soil data for public free access, *International Journal of Digital Earth, 4 (5)*, 434-443.

Ibid, pp. 434-443.

Bhattacharyya, T., Sarkar, D., Gangopadhyay, S.K., Dubey, P.N., Baruah, U., Chamuah, G.S., Mukhopadhyay, S., Nayak, D.C., Maji, A.K., Saxena, R.K., Barthwal, A.K., Krishna, N.D.R., Mandal, C. and Sahgal, J. 1998. Soils of Tripura.II- Suitability for Rubber, *Agropedology, 8*, pp. 55-60.

Chattopadhyay, S. 2015. Environmental Consequences of Rubber Plantations in Kerala, *NRPDD, Discussion Paper No. 44*, Centre for Development Studies, Thiruvananthapuram, pp. 1-53.

4
Role of Different Agencies on the Growth and Distribution of Rubber Plantation in Tripura

4.1 Introduction

Rubber plantation in Tripura is a success, and the previous chapter suggests that area under rubber plantation has grown tremendously. The distribution of rubber plantation suggests that it has undergone a change and at present mainly concentrated in West Tripura and South Tripura districts. Different agencies have played a distinctive role in the growth and distribution of rubber plantation in Tripura which has crossed fifty years since its inception in 1963. As already mentioned, Tripura is at present the second largest producer of NR in India and also has large area under rubber plantation ranking second after Kerala. Rubber plantation here has met much resistance initially from the local people but through policy and institutional interventions mainly conceived and implemented by the Rubber Board and the different agencies functioning under the state government, rubber plantation have proved to be an agent of change in this state ushering in the much needed socio- economic development of the *jhumias* and the marginalized landless population (Bhowmik, 2006; Joseph et al., 2010; Sharma et al., 2011; Sharma, George K and Dey, 2014). At present there are three main agencies looking after the rehabilitation of the *jhumias* and landless population through rubber plantation. This calls for an understanding of the activities executed by different agencies viz. TFDPCL and TRPCL (state government undertaking) and the Rubber Board (Ministry of Commerce and Industry, Government of India) in promoting the growth and distribution of rubber plantation in the state as well as rehabilitating the marginalized communities living here.

4.2 Role of different rubber growing agencies in Tripura

At present there are three major rubber growing agencies working in the state. Consequently, the role of these agencies in the growth and distribution of area under rubber plantation in the state are discussed which are as follows:-

4.2.1 Tripura Forest Development and Plantation Corporation Limited (TFDPCL)

The TFDPCL was established in 1976 mainly with the objective to take advantage of the institutional financing in the wake of the National Commission of Agriculture recommendation to increase the productivity of forest through human intervention and such forests are also to have commercial significance. Consequently, the approach to forestry development includes not only to increase forest area but also to introduce such trees which can be of commercial significance (Das, 2014) hence plantation of rubber trees are encouraged in the state.

The TFDPCL is a public-sector organization functioning under the government of Tripura. TFDPCL was established on the 26th of March 1976 under the Companies Act 1956. TFDPCL was set up with the objective to achieve sustainable development by realizing the *Triple-Bottom-Line* viz. to achieve economic viability, create socially desirability towards rubber plantation and create conditions for the sound management of the environment in the state (TFDPCL, 2016a). TFDPCL functions under the administrative control of the Tripura Forest Department (DESPD, 2016). It is to be mentioned here that the Forest Department has stopped its activities of expansion of area under rubber plantation after 1975. Consequently, the area under rubber plantation developed by the Forest Department from 1963 to 1975 has been transferred to the TFDPCL and thus inherited by the TFDPCL (Bhowmik, 2006; Viswanathan and Bhowmik, 2014; TFDPCL, 2016a).

4.2.1.1 Aims of the TFDPCL

The TFDPCL (TFDPCL, 2016a) targets are to meet seven objectives which are as follows:-

i) Afforestation of degraded forestlands through rubber plantation,

ii) Economically rehabilitate the indigenous *jhumias*, marginalized section of ST and SC families along with other poor rural families through rubber plantation by implementing welfare schemes provided by the centre and the state governments,

iii) Managing rubber plantation as well as help in the production of raw rubber into marketable form and also production of high quality rubber-wood for further value addition,

iv) Promote eco-friendly rubber-wood products and reduce pressure on natural forest,

v) Promote the establishment of rubber latex and rubber-wood based industries for value addition,

vi) Encourage use of organic manures in the plantation, and

vii) Improve the socio-economic conditions of the communities who depend on rubber plantation by encouraging responsible management practices, help them to adapt to climate change and to achieve environmental benefits like carbon sequestration, soil conservation as well as adopt measures to protect soil moisture.

4.2.1.2 Functioning of the TFDPCL

The TFDPCL on an experimental basis started their project of planting rubber trees on degraded forestland and simultaneously helping in weaning away the practice of *jhum* cultivation of one hundred indigenous tribal families in Warangbari village located in West Tripura district. The TFDPCL officials introduced wage employment for the *jhumias* in the rubber plantation here as rubber plantation has a gestation period of about seven years before the plantation are of commercial significance. However, in the initial stage the villagers protested against such plantation as they were worried about the loss of community land for *jhum* cultivation as well as other common pool natural resources essential for carrying on their livelihood practices. The tribes here live in close proximity to nature where their livelihood practices are intricately related to their natural surroundings providing environmental income and sustenance for the marginalized. The officials of TFDPCL were keen on implementing this project, hence they came-up with a solution, where the villagers were promised that on attaining maturity the rubber plantation would be allotted to them (Sinha, 2012). Consequently, from landless population they would have better access to land (but cannot own it as the land is leased), this appealed to the beneficiaries of this programme. Consequently, the first rubber plantation-based rehabilitation project was introduced in 1976 at Warrangbari village. The TFDPCL emerged as the pioneer in successful implementation of the first rubber plantation- based rehabilitation project by providing 1 ha of rubber plantation to each indigenous *jhumias* family at Warrangbari village (TFDPCL, 2016a). Success of the project (both commercial and rehabilitation programme) has not only encouraged the planners but has also generated trust and confidence amongst the poor ST inhabiting mainly the hills and uplands (Das, 2014).

Under this model the members of the rehabilitated families initially work as daily wage labourers and later earn enjoy income from the sale of latex when the rubber trees start yielding latex (TRPCL, 2012 and 2016). As per this model, rubber plantations are raised on the allotted land of 1 ha per family, and various benefits and skill development are provided by the TFDPCL to the beneficiaries for raising rubber trees. After maturity of rubber trees, the beneficiaries collect the latex and bring it to the resettlement centre, where it is processed into rubber sheet, dried and smoked by the beneficiaries. The TFDPCL then procures the rubber sheets and scrap at the prevailing market rates and pursues further marketing operations (Viswanathan and Bhowmik, 2014).

It is interesting to note that no customary rights are given to the beneficiaries over the plantation area other than the TFDPCL (ownership of land remains under the Corporation) as these areas are given by the state government on a long-term lease to the TFDPCL (TFDPCL, 2016a).

Besides supervising the beneficiary members and workers (involved in commercial rubber plantation) for raising rubber trees, the TFDPCL also imparts training on tapping and processing of latex into rubber sheet to the beneficiaries prior to their engagement, so that they are well conversant with various aspect of rubber trees for tapping and processing of latex up-to coagulation. The TFDPCL also conducts refresher training for the tappers to upgrade their skill with new techniques and tapping knives launched by the Rubber Board from time to time. Processing workers are trained mainly to handle sheet rolling machines, proper sun drying and smoke drying, further they are trained to monitor the common defects noticed in smoked dried sheets, knowledge regarding grading of sheets mainly RSS-4 and RSS- 5, packing and baling of smoke dried rubber sheets, and marking each Bale with RPC- Code and rubber Grade. The TFDPCL also provides safety measures for the tappers and processing workers like providing them with hunting shoes and woollen jerseys, gum boot/apron/musk to the processing workers, First Aid Box at each RPC, rest sheds for the shelter for tappers during rains, maintenance of base line cleaning during winter to avoid fire hazards, approach path/foot path/steps, helpers are also paid overtime dues for carrying excess latex, tappers are further motivated to attend health camps organized by civil administration at nearby places, chemicals/insecticides/fungicides are kept carefully at each RPC, telephone numbers (mainly of police station, fire service, hospital, ambulance), drinking water facilities at each RPC for tappers and workers are also provided, recreation and entertainment (like carom board and volley ball etc), as well as other amenities are provided to the tappers/workers, safety measures for felling of trees and conversion into logs, reimbursements of medical bills for workers etc are provided by the TFDPCL (TFDPCL, 2016a). Consequently, a lot of welfare measures, skill development, and economic benefits associated with rubber plantation under TFDPCL have encouraged more marginalized population to join in this model of development. Accordingly, on the 16th of February 1981, the government of Tripura signed an agreement with TFDPCL to develop 5,200 ha of forestland for cultivation of rubber trees, citronella and bamboo on a long-term lease of 42 years (TFDPCL, 2016a).

4.2.1.3 Growth of rubber plantation under the TFDPCL

As the success of TFDPCL is high another 9,019.52 ha of degraded forestland has been transferred to the TFDPCL for the purpose of rubber plantation. In 1997 the Minister of Environment and Forest, Government of India, approved 1,500 ha of land for resettling of the *jhumias* through rubber plantation (this project took off in 1998- 99 and the funding was provided by the government of Tripura). In 2005 another 3,373.79 ha of degraded forestland located along the Indo-Bangladesh International border was transferred to TFDPCL for setting up rubber plantation (TFDPCL, 2016a).

4.2.1.4 TFDPCL and the emerging pattern of increase of area under rubber plantation

4.2.1.4.1 Share of area under rubber plantation developed by the TFDPCL to the total area under rubber plantation of the state

The share of area under rubber plantation developed by the TFDPCL to the total area under rubber plantation of the state in its initial year (1976-77) was as high as 25.78 percent but at present its share of area under rubber plantation has fallen to 15.64 percent (appendix 4). The average share of area under rubber plantation developed by the TFDPCL to the total area under rubber plantation in the state from 1976-77 to 2015-16 is to the tune of 44.35 percent.

As depicted in figure 4.1, the share of area under rubber plantation developed by the TFDPCL to the total area under rubber plantation of the state started increasing from 1976-77 reaching to its peak during 1981-82 when the share of area under rubber plantation developed by the TFDPCL was as high as 82.12 percent (appendix 4) of the state's total area under rubber plantation. After 1982 the share of area under rubber plantation developed by the TFDPCL to the total area under rubber plantation of the state started declining continuously and the lowest share of area under rubber plantation of this agency to the total area under rubber plantation of the state is during 2015-16 with only 15.64 percent (appendix 4). Consequently, the share of area under rubber plantation of the TFDPCL to the total area under rubber plantation of the state is declining as other agencies and private owners have come up bringing more area under rubber plantation in the state.

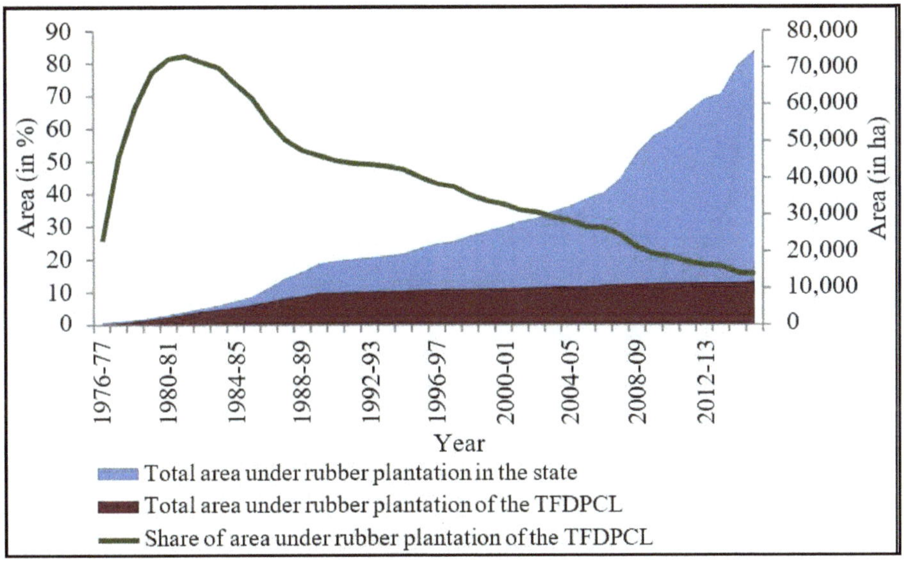

Fig. 4.1: Share of area under rubber plantation developed by the TFDPCL to the total area under rubber plantation of the state (19976-77 to 2015-16).

4.2.1.4.2 Role of TFDPCL in the expansion of area under rubber plantation

The total area under rubber plantation developed by the TFDPCL during the initial year (1976-77) was only 148 ha which at present has increased to 11,624 ha (appendix 5) with an overall growth rate as high as 7,754.05 percent in about four decades. Figure 4.2 clearly suggests the rapid increase of area under rubber plantation developed by the TFDPCL from its inception till 1989-90 but after 1990, the expansion of area under rubber plantation developed by the TFDPCL is found to be increasing at a slower pace. The drastic increased in area under rubber plantation developed by the TFDPCL till 1989-90 may be due to extension of area under rubber plantation on the availability of favorable degraded land (caused mainly by *jhum* cultivation practices) for restoring through commercial rubber plantation and also implementation of rehabilitation programme through rubber plantation (for economically and socially marginalized section of peoples) as well as acceptance of rubber plantation by the rural poor populace. Besides success of commercial rubber plantation, success of the rehabilitation programme has not only encouraged the planners for further extension of rubber plantation but has also generated trust and confidence amongst the poor tribes inhabiting the hills and uplands (Das, 2014). On the other hand, the steady increased in extension of rubber plantation area may be due to minimizing land availability for rubber plantation for the TFDPCL by the state government as other agencies (TRPCL, Rubber Board and TTAADC and SDM) and private owners has also started rubber plantation on available favorable land for rubber plantation.

To show the trend in the expansion of area under rubber plantation developed by the TFDPCL, the annual growth rate of area under rubber plantation has been computed which is depicted in figure 4.2.

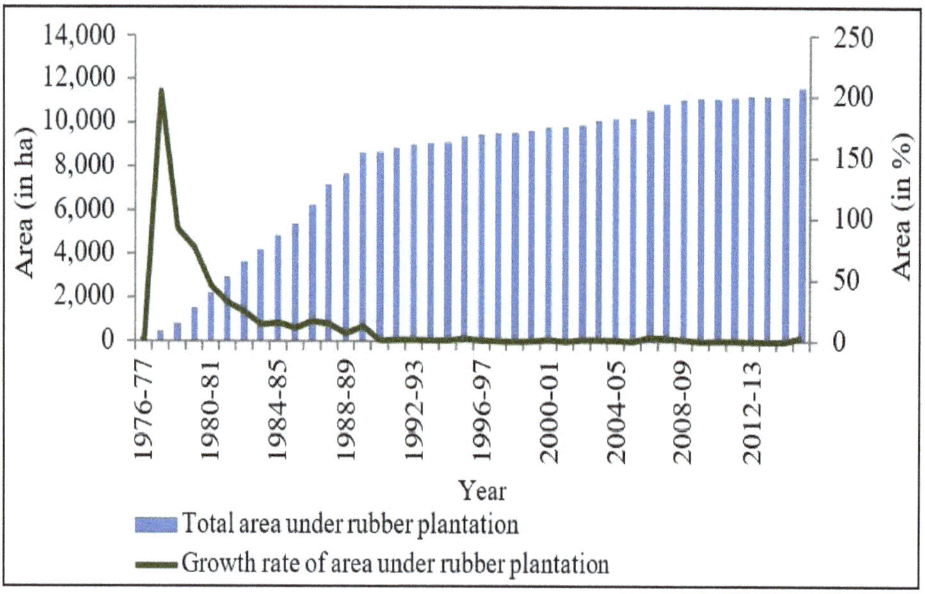

Fig. 4.2: Growth rate of area under rubber plantation developed by the TFDPCL (19976-77 to 2015-16).

It is interesting to note that in the initial year the annual growth rate of area under rubber plantation is as high as 204.73 percent (1977-78) which fell below 100 percent in the next year (1978-79). It is interesting to note that from 1990-91 to 2015-16 the area under rubber plantation developed by the TFDPCL has been increasing with the annual growth rate of below 4 percent (figure 4.2 and appendix 5).

The annual growth rate of area under rubber plantation developed by the TFDCL indicates that from 1976-77 to 2015-16 the average rate of growth in area under rubber plantation developed by this agency is about 15.32 percent. Further analysis of the area under rubber plantation developed by the TFDPCL suggests that it is increasing continuously but at a declining annual growth rate (figure 4.2).

As the TFDPCL has completed about forty years (1976-77 to 2015-16) in the development of area under rubber plantation in the state, the decadal wise compound annual growth rate (CAGR) has also been looked into (table 4.1).

Table 4.1: Decadal CAGR of rubber plantation area of TFDPCL

Decadal year (DY)	Area under rubber plantation (in ha)	CAGR of area under rubber plantation (in %)
1976-77 to 1985-86 (1st DY)	5,361	43.19
1986-87 to 1995-96 (2nd DY)	4,029	4.13
1996-97 to 2005-06 (3rd DY)	824	0.75
2006-07 to 2015-16 (4th DY)	1,410	0.93

Source: Decadal wise CAGR computed by the researcher.

It is observed that in the initial decade (1976-77 to 1985-86), area under rubber plantation developed by the TFDPCLwas as high as 5,361 ha with a CAGR of 43.19 percent suggesting a faster CAGR in comparison to the next decade. In the second decade (1986-87 to 1995-96), the TFDPCL developed 4,029 ha of rubber plantation with a CAGR as low as 4.13 percent suggesting a sharp decline in CAGR in comparison to the previous decade. In the third decade (1996-97 to 2005-06), area under rubber plantation developed by the TFDPCL was as low as 824 ha with a still lower CAGR of 0.75 percent suggesting a further declining CAGR in comparison to the previous decade. In the fourth decade (2006-07 to 2015-16), area under rubber plantation developed by the TFDPCL was relatively high (1,410 ha) with a CAGR of 0.93 percent suggesting a slight increase of CAGR in comparison to the previous decade. These suggest that the decade wise increase in area under rubber plantation developed by the TFDPCL has increased at declining rate till the third decade. In the fourth decade however, the area under rubber plantation developed by the TFDPCL has increased substantially from about 824 ha to 1,410 ha suggesting a gain of about 586 ha within a span of 10 years.

4.2.1.4.3 Role of TFDPCL in the annual gain of area under rubber plantation

The highest gain of area under rubber plantation (974 ha) developed by the TFDPCL is observed during the year 1989-90 and the least gain of area under rubber plantation is marked in the year 2013-14 having a gain of just 2 ha (appendix 5). The high annual gain in area under rubber plantation developed by the TFDPCL from the initial phase till 1989-90 can be attributed to the availability of suitable land for rubber plantation, besides financial assistance by the state government and targets initiated for commercial rubber plantation and rehabilitation programme through rubber plantation.

From 1976-77 to 2015-16, the gain of area under rubber plantation developed by the TFDPCL has been increasing at an average of 290.60 ha. The standard deviation of annual gain in area under rubber plantation developed by the TFDPCL from 1976-77 to 2015-16 is 290.14 and the coefficient of variation is to the tune of

99.84. This suggests that there is a large variation in the annual gain of area under rubber plantation developed by the TFDPCL from 1976-77 to 2015-16.

The annual gain in area under rubber plantation developed by the TFDPCL from 1976-77 to 2015-16 range from 2 ha (lowest) to 974 ha (highest). Consequently, the nature of gain in area under rubber plantation developed by the TFDCPL has been grouped into five categories as shown in table 4.2.

Table 4.2: Nature of gain in area under rubber plantation and number of years (developed by the TFDPCL from 1976-77 to 2015-16)

Nature of gain	Area under rubber plantation (in ha)	No. of year
Very high gain	Above 800	3
High gain	601 – 800	5
Moderate gain	401 – 600	5
Low gain	201 – 400	4
Very low gain	below 200	23

Source: Computed by the researcher.

It is observed that for most of the year (twenty three numbers of years) the TFDPCL has developed rubber plantation of below 200 ha per year while in fourteen numbers of year the TFDPCL has developed rubber plantation of above 200 ha per year. In three number of years i.e. during 1986-87, 1987-88 and 1989-90, TFDPCL has developed more area under rubber plantation (above 800 ha per year). High gain in area under rubber plantation ranging 601-800 ha per year is observed during 1979- 80, 1980-81, 1981-82, 1982-83 and during 1984-85. Another five numbers of years i.e. 1978-79, 1983-84, 1985-86, 1988-89 and 2015-16 are suggestive of having moderate gain in area under rubber plantation ranging 401-600 ha per year. Low gain in area under rubber plantation ranging 201-400 ha per year are observed in four numbers of years i.e. 1977-78, 1995-96, 2006-07 and 2007-08. The lowest gain of area under rubber plantation below 200 ha per year are marked for the maximum years i.e. twenty three numbers of years starting in 1976-77, 1990-91, 1991-92, 1992-93, 1993-94, 1994-95, 1996-97, 1997-98, 1998-99, 1999-00, 2000-01, 2001-02, 2002-03, 2003-04, 2004-05, 2005-06, 2008-09, 2009-10, 2010-11, 2011-12, 2012-13, 2013-14 and during 2014-15 (table 4.2).

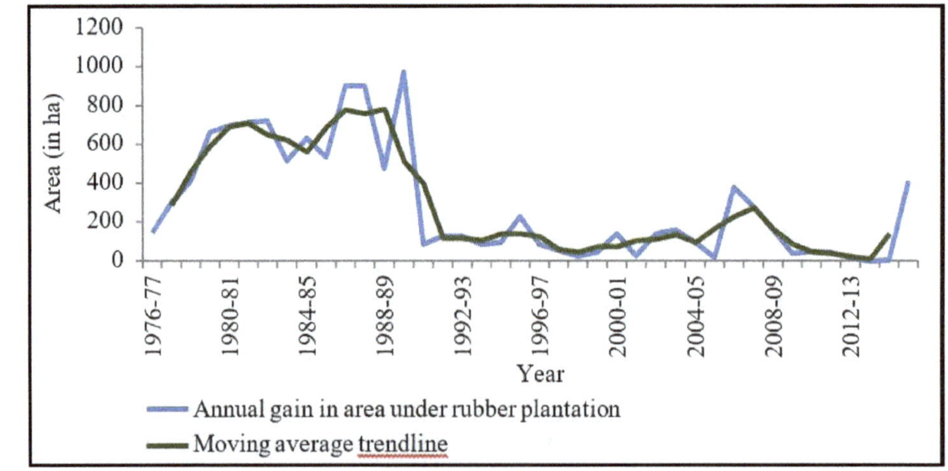

Fig. 4.3: Trend of annual gain in area under rubber plantation developed by the TFDPCL from 1976-77 to 2015-16 (applying 3 year moving average).

If a trend line (taking 3 yearly moving average) is fitted into the annual gain in area under rubber plantation developed by the TFDPCL (figure 4.3), it indicates that the area under rubber plantation has been increasing significantly from the year 1976- 77 reaching its peak in the year 1981-82. After 1982, the trend in the gain in area under rubber plantation started declining till 1984-85. After 1985 the gain in area again started increasing significantly till 1986-87 but again declined in the following year of 1987-88.

Subsequently in the year of 1988-89 the gain in area under rubber plantation developed by the TFDPCL suggests a steep gain reaching its second peak where the gain of area to the previous year is to the tune of 785 ha. After 1989, the gain in area under rubber plantation developed by the TFDPCL indicates a declining trend till 1991-92 which continues till 2015-16 with a fluctuating trend where one peak is noticed in the year of 2007-08 (appendix 5). The trend line suggests that there are three peaks in the gain of area under rubber plantation developed by the TFDPCL where the first peak is marked in the year of 1981-82, followed by 1988-89, and the last peak is marked in the year 2007-08.

4.2.1.5 Distribution of area under rubber plantation developed by the TFDPCL

For district wise distribution of area under rubber plantation developed by TFDPCL has been analyzed from the data of 2014-15. Table 4.3 reveals that till 2014- 15 TFDPCL had developed about 11,211 ha of area under rubber plantation in 67 RPCs. It is observed that TFDPCL has developed rubber plantation only in three districts of the state viz. North Tripura, West Tripura, and South Tripura districts. Maximum area under rubber plantation developed by TFDPCL is noticed in South Tripura district (44.48 %) where 16 RPCs are located. Next is West Tripura district where the share of area under rubber plantation is 30.88 percent. Here, 25 RPCs are located. The least share of area under rubber plantation to the total area under rubber plantation developed by the TFDPCL is in the North Tripura district accounting for only about a quarter share (24.64 percent). Here there are only 16 RPCs. It is observed that the TFDPCL does not have its presence in Dhalai district in spite of the fact that it is in this district that the first rubber plantation was introduced.

Table 4.3: District wise distribution of area under rubber plantation developed by the TFDPCL (2014-15)

District	Area under rubber plantation (in ha)	Share in %	RPC (no.)	Share in %
North Tripura	2,762	24.64	16	23.88
Dhalai	-------	-------	---	-------
West Tripura	3,462	30.88	25	37.31
South Tripura	4,987	44.48	26	38.81
Total	11,211	100	67	100

Source: TFDPCL, 2016.

In regards to district wise share of area under rubber plantation developed by the TFDPCL to the district wise total area under rubber plantation of the state during 2014-15 as depicted in table 4.4, TFDPCL has the highest share of area under rubber plantation in North Tripura district (29.49 %) followed by

South Tripura district (22.94 %) and the least contribution is in West Tripura district with 9.80 percent share. It is observed that although the TFDPCL has developed more area under rubber plantation in West Tripura and South Tripura districts than the North Tripura district but its contribution of area under rubber plantation to the total area under rubber plantation of West Tripura and South Tripura districts are less than the North Tripura district. The reason may be due to large contribution of area under rubber plantation by private and individual owners in West Tripura and South Tripura districts than the North Tripura district.

Table 4.4: District wise share of area under rubber plantation developed by the TFDPCL to the district wise total area under rubber plantation of the state (2014-15)

District	Area under rubber plantation (in ha)	Share in %	RPC (no.)	Share in %
North Tripura	2,762	24.64	16	23.88
Dhalai	-------	-------	---	-------
West Tripura	3,462	30.88	25	37.31
South Tripura	4,987	44.48	26	38.81
Total	11,211	100	67	100

Source: Rubber Board, 2016; TFDPCL, 2016 (share of area under rubber plantation of TFDPCL computed by the researcher).

4.2.1.5.1 District wise concentration of area under rubber plantation developed by the TFDPCL (2014-15)

In order to measure the district wise concentration of area under rubber plantation developed by the TFDPCL, concentration index (CI) score has been computed which is further mapped as shown in figure 4.4. The CI score value of the different districts ranges from 1.14 to 0.81. Consequently, the concentration of area under rubber plantation has been categorized into high concentration, moderate concentration and low concentration.

It is noticed from figure 4.4 that South Tripura district has the highest concentration of area under rubber plantation with a CI score of 1.14. North Tripura district has moderate concentration of area under rubber plantation with a CI score of 0.95 while West Tripura district has the lowest concentration of area under rubber plantation with a CI score of 0.81.

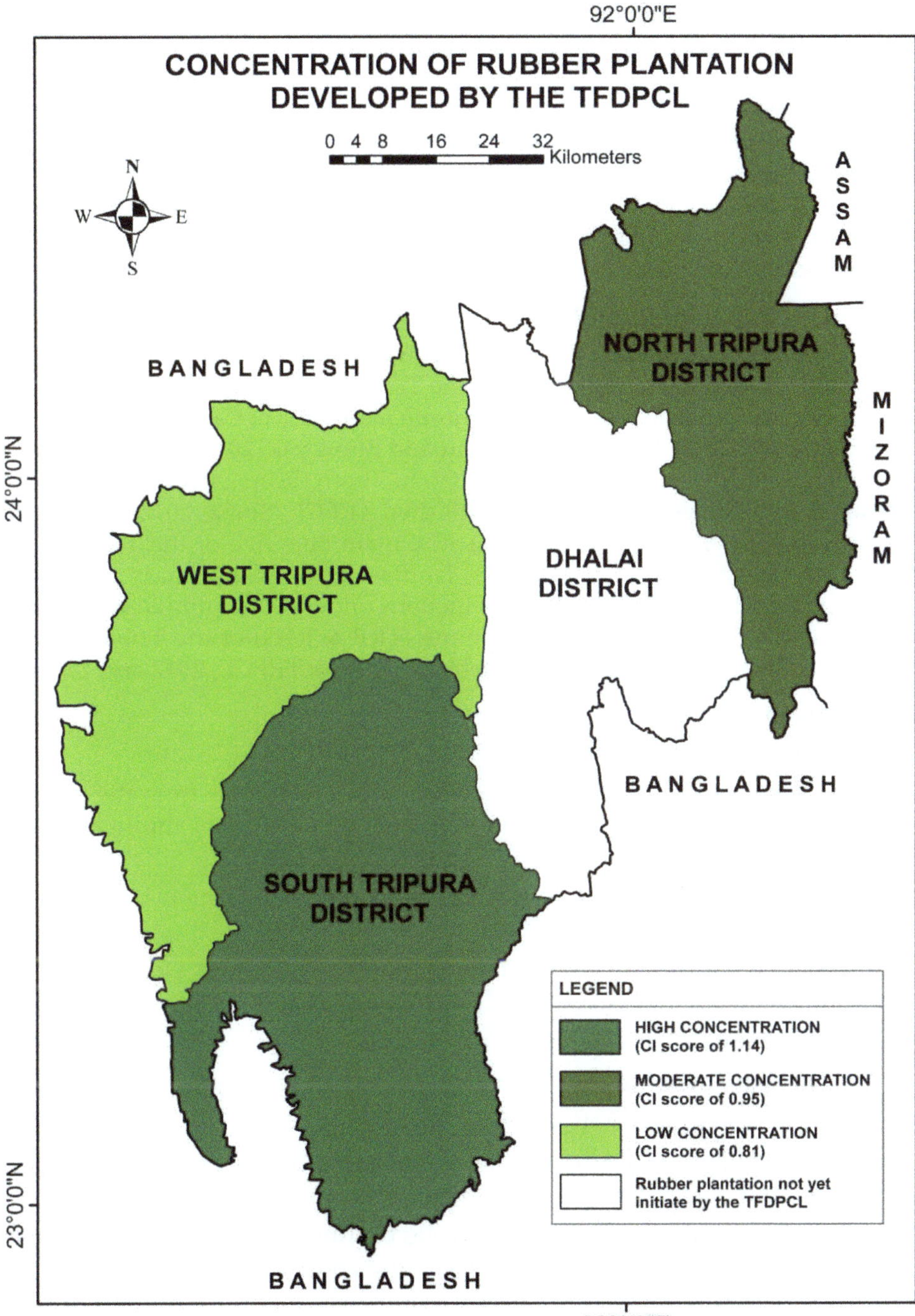

Fig. 4.4: District wise concentration of area under rubber plantation developed by the TFDPCL (2014-15).

4.2.1.6 Level of achievement of the basic objectives of the TFDPCL

As already mentioned, TFDPCL has adopted rehabilitation of degraded forestland through commercial rubber plantation as its primary objective along with rehabilitation of ST *jhumias* in the state. The commercial rubber plantation is termed as 'TFDPC Plantations' while the resettlement project through rubber plantation is termed as 'Welfare Plantations'. Of the 11,211 ha of rubber plantation developed by the TFDPCL distributed in the 67 RPCs, 8,118 ha (72.41%) is for commercial purpose distributed in 38 RPCs (56.72%) while only 3,093 ha (27.59%) is for resettlement purposes distributed in 29 RPCs (43.28%) (TFDPCL, 2016a).

4.2.2 Tripura Rehabilitation Plantation Corporation Limited (TRPCL)

TRPCL was established on the 3rd of February 1983 (Debbarma, 2014; TRPCL, 2012 and 2016), about 10 years later than the TDPCL. The rapid expansion of area under rubber plantation in the state due to implementation of Case Subsidy Scheme by the Rubber Board in 1980 (Viswanathan and Bhowmik, 2014), and the success of rehabilitation programme by the TFDPCL has been the main reasons that have encouraged the state government to established TRPCL (Sinha, 2012; Debbarma, 2014; Viswanathan and Bhowmik, 2014). The main objective of the TRPCL is to economically rehabilitate the ST *jhumias*, landless and small farmers principally through rubber plantation. The TRPCL functions under the Tripura Rehabilitation Plantation and Primitive Group Programme (TRP & PTG) of the Tribal Welfare Department, Government of Tripura (Debbarma, 2014; TRPCL, 2012 and 2016).

4.2.2.1 Aims of the TRPCL

The TRPCL prime objectives (Debbarma, 2014; TRPCL, 2016) are-

- i) To economically rehabilitate the *jhumias* (wean *jhumias* from the practice of *jhum* cultivation), tribal landless and small farmers through rubber plantation,
- ii) To develop the economic condition of the marginal farmers by providing them livelihood options through rubber plantation,
- iii) Function as an extension wing of the Tribal Welfare Department for implementation of government policies and programmes,
- iv) To generate job facilities in the rural areas,
- v) To produce genuine planting materials,
- vi) Production of rubber sheets and its marketing,
- vii) Intercropping of legumes and tuber crops under immature rubber plantation for food security and income during gestation period,
- viii) Shelterbelt for rubber with tree species bearing edibles e.g. tree beans and commercially important species like Agar and bamboo,
- ix) To provides training and capacity building both for the official staffs and its beneficiaries, and
- x) To initiate formation of Welfare Fund for micro-credit with appropriate guidelines.

4.2.2.2 Functioning of the TRPCL

For the rehabilitation programme, *firstly*, the list of selected families/ beneficiaries and particulars of land for rubber plantation duly endorsed by the ADC Village Councils/Panchayat Samitis or Block Advisory Committees are forwarded by the BDOs to the concerned Zonal Managers. After this process, the TRPCL initiates cultivation of rubber saplings and manages it with fund provided by the Tribal Welfare Department by directly involving the concerned beneficiaries as wage laborer till the plants gain maturity i.e. six to seven years. Here, an annual target for rubber plantation is fixed by the Tribal Welfare Department, after consultation with the TRPCL. Initially the TRPCL cultivated rubber trees on 1.5 ha per family, but this could not continue further and now 1 ha of rubber plantation per family is the standard norm (TRPCL, 2012; Debbarma, 2014). However, this may vary from place to place on availability of land. It is to mention here that apart from rehabilitating the *jhumias*, tribal landless and small farmers, the TRPCL has also initiated rehabilitating the surrendered extremists after the peace accord between the state government and the Tribal National Volunteers (TNV) in 1988-89 (Viswanathan and Bhowmik, 2014).

During the initial period of rehabilitation programme of the TRPCL, lack of confidence was prevalent among the beneficiaries on the prospect of rubber plantation as a viable livelihood option. Consequently, some beneficiaries started abandoning the plantations (back to *jhum* cultivation) as a result of long gestation period and poor management and protection of the plantation from grazing, fire etc. Apart from this natural hazards like storm and cyclones have destroyed and damaged rubber plantation. Poor accessibility to some rural areas on the other hand has increased difficulties for the TRPCL to maintain and manage the rubber plantation (TRPCL, 2012).

To counter these problems of long gestation period where income flow is restricted, the TRPCL motivated the beneficiaries to raise Agar plantation as shelterbelt for rubber plantation and homestead under the scheme of National Medicinal Plants Board so to ensure flow of income for beneficiaries during the long gestation period (TRPCL, 2012). Bamboo plantation has also been raised as shelterbelt with financial assistance from Tripura Bamboo Mission from 2008-09 to 2009-10 (TRPCL, 2012; Debbarma, 2014). The TRPCL also has raised 1 ha of rubber plantation with arecanut as a trial at Panbagan RPC in North Tripura district. Inter- cropping is also encouraged; some of the beneficiaries under Gandacherra RPC have cultivated sugarcane, arhar and cow pea as inter-crop with rubber trees in 2014 which has been a success (Debbarma, 2014).

The TRPCL also signed an agreement with each beneficiary (TRPCl, 2012; Debbarma, 2014) for the following:-

i) To ensure the beneficiaries do not transfer the rubber plantation land or lease out the plantation to others,

ii) To ensure the beneficiaries sell the latex to the TRPCL for a fair price determined by the Board of Director of the TRPCL where the price is revised from time to time according to the prevailing market price and the cost of processing and other overheads.

The TRPCL imparts guidance and supervision about maintaining the rubber trees to the beneficiaries so to enable them to maintain and take up the matured rubber plantation. The officer-in-charge of each RPC of the TRPCL also associates and involves the beneficiaries through regular meetings and encourages members of the beneficiary households to be part of the Beneficiary Committee/Rubber Producers Committee in each RPC so that they can share their needs and ideas regarding rubber plantation with the TRPCL. The TRPCL also organises extensive training on tapping and processing, marketing and also on first aid, use of solar disinfected water (SODIS), bud-grafting, mushroom cultivation, bee keeping etc both for the official staffs and its beneficiaries for appropriate cultivation and management of rubber plantation and also for all round development of the target group (TRPCL, 2012; Debbarma, 2014).

During 2010-11, some of the training and capacity development activities organised by the TRPCL (TRPCL, 2012) are-

i) five days residential training on rubber cultivation and management at NRECT where 93 persons are selected including beneficiaries and field staff of the TRPCL,

ii) field training on bud grafting, tapping and processing for 173 beneficiaries and field staff of the TRPCL has also been organised, and

iii) The TRPCL has also organised inter Zonal exposure visit for 24 beneficiaries.

Apart from the already mentioned roles and activities, the TRPCL has also undertaken some new initiatives (TRPCL, 2012; Debbarma, 2014) which are as follow:-

i) Residential training of the TRPCL staff on rubber related and other subjects e.g. mercantile accounting system, micro credit, health insurance, moral values, sports and cultural activities, etc.,

ii) Establishment of central nursery to meet the growing demand in the state,

iii) Formation of welfare fund for micro-credit with appropriate guidelines,

iv) Provision for computer training to TRPCL employees through National Institute of Technology during office hours,

v) Introduction of tree beans (Parkia timoriana*), sajna, bakful, keli kadam, satkara* (citrus macroptera) etc edibles in homestead land of beneficiaries to meet their nutritional needs and for incremental income,

vi) Promotion of intercropping of legumes and tuber crops in the immature rubber plantation for food security and income during gestation period,

vii) Motivating the beneficiaries at Dhanbilash RPC in North Tripura district to run the first self-financing coaching class for students from the beneficiary families; this to be replicated in some of the other centres,

viii) Initiation of one Research and Development project with the Rubber Research Institute of India (RRII) for trail of winged bean as an alternative crop covers under rubber and edible tubers as an additional crop,

ix) Supply of raw latex to industries instead of conversion to sheets to avoid destruction of tree cover in the countryside,

x) Attempt to make 100 percent payment through individual bank account of beneficiaries, and

xi) Project taken up at Surendranagar with the assistance of Tribal Welfare Department.

Consequently, the tribal beneficiaries are serious and are considering the rubber plantation as important income source and are also appreciating the value of their property i.e. rubber plantation land (TRPCL, 2012; Das, 2014; Debbarma, 2014).

4.2.2.3 TRPCL and the emerging pattern of increase of area under rubber plantation

4.2.2.3.1 Share of area under rubber plantation developed by the TRPCL to the total area under rubber plantation of the state

The share of area under rubber plantation developed by the TRPCL to the total area under rubber plantation of the state during its inception (1976-77) was just 1.15 percent and during 2015-16 has a share of 12.31 percent (appendix 6). Thus, its role as an agency looking after rubber plantation of the state is second to that of TFDPCL. The average share of area under rubber plantation developed by the TRPCL to the total area under rubber plantation of the state (1984-85 to 2015-16) is to the tune of 13.19 percent.

The trend of share of area under rubber plantation developed by the TRPCL to the total area under rubber plantation of the state is shown in figure 4.5. The figure indicates that the share of area under rubber plantation developed by the TRPCL to the total area under rubber plantation of the state started increasing significantly from 1984-85 till the year 1989-90. But after 1990, the share of area under rubber plantation developed by the TRPCL suggests an increasing trend till 1994-95 but at a slower pace. From 1995-96 to 1998-99, the share of area under rubber plantation suggest a fluctuating trend. After 1999, the share of area under rubber plantation developed by the TRPCL to the total area under rubber plantation of the state started increasing significantly reaching its maximum share (17 percent) during the year 2006-07. After 2007, the share of area under rubber plantation developed by TRPCL to the total area under rubber plantation of the state is declining each year till 2015-16. It is noticed that TRPCL has maintained above 10 percent share of area under rubber plantation from 1988-89 to 2015-16 (appendix 6) although in the recent years the share has been declining as other agencies, private sector and individual owners have come up bringing more area under rubber plantation in the state.

4.2.2.3.2 Role of TRPCL in the expansion of area under rubber plantation

The total area under rubber plantation developed by the TRPCL during the initial year (1984-85) was only 75 ha which has increased to 9,154 ha during 2015-16 (appendix 6) with an overall growth rate of 12,105.33 percent in about slightly above three decades. Figure 4.6 clearly suggest a rapid expansion of area under rubber plantation developed by the TRPCL from its inception till 1989-90 but after

1990, the expansion of area under rubber plantation developed by the TRPCL has been increasing but at a steady nature.

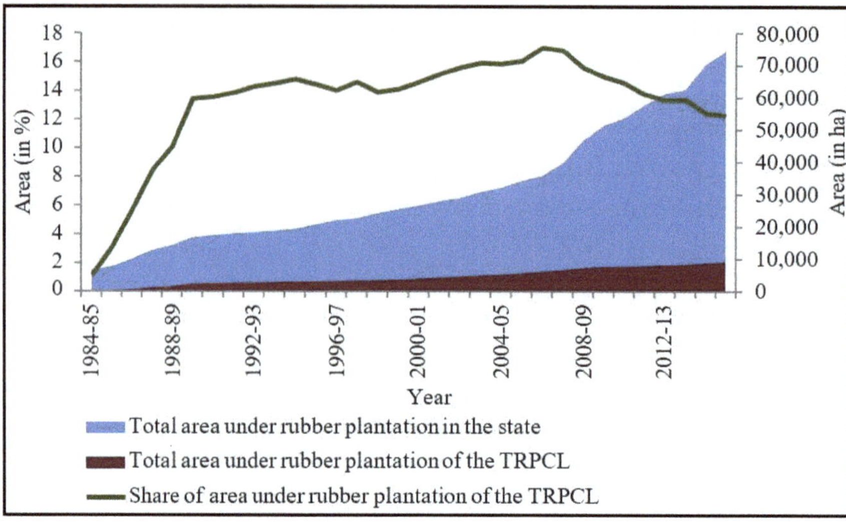

Fig. 4.5: Share of area under rubber plantation developed by the TRPCL to the total area under rubber plantation of the state (1984-85 to 2015-16).

The continuous increased in area under rubber plantation developed by the TRPCL is the support of the state government for extension of rubber plantation through rubber plantation-based economic-rehabilitation programme for the economically and socially marginalized section of peoples and the result of acceptance of the rehabilitation programme by the poor ST population throughout the state; as the main objective of the TRPCL is to economically rehabilitate the *jhumias,* landless, and small farmers mainly belonging to ST population inhabiting rural areas principally through rubber plantation (TRPCL, 2016).

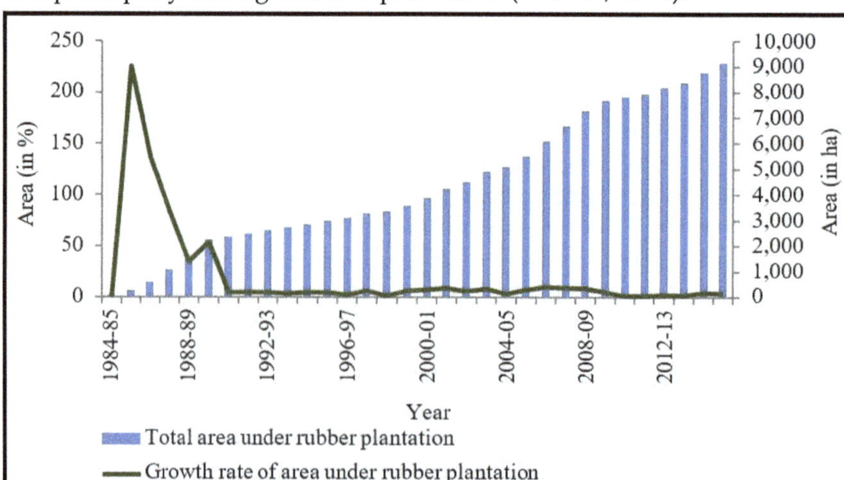

Fig. 4.6: Growth rate of area under rubber plantation developed by the TRPCL (1984-85 to 2015-16).

To show the trend in the expansion of area under rubber plantation developed by the TRPCL, the annual growth of area under rubber plantation has been computed which is depicted in figure 4.6.

It is observed from figure 4.6 that area under rubber plantation developed by the TRPCL is continuously increasing but at a declining annual growth rate. The highest annual growth rate of area under rubber plantation developed by the TRPCL has been observed in the year 1985-86 when the annual growth rate was as high as 226.67 percent while the lowest annual growth rate has been observed in the year 2010-11 recording an annual growth rate as low as 1.65 percent only. It is also observed that from 1990-91 onwards the annual growth rate of area under rubber plantation developed by the TRPCL is below 10 percent except in the year 2006-07 when the annual growth rate is found to be marginally higher about 11 percent (appendix 7). The average growth rate in area under rubber plantation developed by the TFDCL from 1977-78 to 2015-16 is to the tune of 22.04 percent. TRPCL from its inception (1984-85 to 2015-16) has completed almost three decades of functioning in Tripura, thus the decadal wise compound annual growth rate (CAGR) has also been looked into (table 4.5).

Table 4.5: Decadal wise CAGR of rubber plantation area developed by TRPCL

Decadal year (DY)	Area under rubber plantation (in ha)	CAGR of area under rubber plantation (in %)
1984-85 to 1993-94 (1st DY)	2,711	43.16
1994-95 to 2003-04 (2nd DY)	2,194	5.58
2004-05 to 2013-14 (3rd DY)	3,467	5.09

Source: Decadal wise CAGR computed by the researcher.

It is observed that in the initial decade (1984-85 to 1993-94), area under rubber plantation developed by the TRPCL was about 2,711 ha with a CAGR of 43.16 percent indicating a faster CAGR in comparison to the next decade. In the second decade (1994-95 to 2003-04), area under rubber plantation developed by the TRPCL was 2,194 ha with a CAGR of 5.58 percent suggesting a significant fall in CAGR in comparison to the previous decade. In the third decade (2004-05 to 2013-14), area under rubber plantation developed by the TFDPCL was 3,467 ha recording a CAGR of 5.09 percent suggesting a further declining CAGR when compared to the previous decade. These suggest that the decade wise increase in area under rubber plantation developed by the TRPCL has increased but at declining rate.

4.2.2.3.3 Role of TRPCL in the annual gain in area under rubber plantation

The highest gain of area under rubber plantation developed by the TRPCL is observed in the initial years of 1989-90 which recorded a gain of 787 ha in comparison to the previous year while the least gain in area under rubber plantation developed by this agency was during the year 1998-99 when the area under rubber plantation increased by only 59 ha in comparison to the previous year (appendix 7).

The high annual gain in area under rubber plantation developed by the TFDCL from 1999-00 to 2008-09 can be attributed to available land for rubber plantation and to reach the target initiated for rehabilitation programme through rubber plantation besides financial assistance by the state government.

The average gain of area under rubber plantation developed by the TRPCL is to the tune of 286.06 ha. The standard deviation of annual gain in area under rubber plantation developed by the TRPCL from 1984-85 to 2015-16 is 183.29 and the coefficient of variation is to the tune of 64.07. This suggests that there is variation in the annual gain of area under rubber plantation developed by the TRPCL from 1984- 85 to 2015-16.

The annual gain in area under rubber plantation developed by the TRPCL from 1984-85 to 2015-16 range from 59 ha (lowest) to 787 ha (highest). Consequently, the nature of gain in area under rubber plantation developed by the TRCPL has been group into five categories as shown in table 4.6. It is interesting to note that very high gain in area under rubber plantation is reported for a period of five years viz. 1987-88, 1989-90, 2006-07, 2007-08, 2008-09 where the gain per year is above 450 ha. Seven years records high gain of area under rubber plantation ranging 351-450 ha per year and the years are 1988-89, 2001-02, 2003-04, 2005-06, 2009-10, 2014-15 and 2015-16. Only three years records a moderate gain of area under rubber plantation ranging 251-350 ha and the years are 1986-87, 200-01 and 2002-03.

Table 4.6: Nature of gain in area under rubber plantation and number of years (developed by the TRPCL from 1984-85 to 2015-16)

Decadal year (DY)	Area under rubber plantation (in ha)	CAGR of area under rubber plantation (in %)
1984-85 to 1993-94 (1st DY)	2,711	43.16
1994-95 to 2003-04 (2nd DY)	2,194	5.58
2004-05 to 2013-14 (3rd DY)	3,467	5.09

Source: Computed by the researcher.

On the other hand six years i.e. 1985-86, 1997-98, 1999-00, 2004-05, 2012-13, and 2013-14 records a low gain of area under rubber plantation ranging 151-250 ha per year, while eleven years i.e. 1984-85,1990-91,1991-92, 1992-93,1993-94, 1994-95, 1995-96, 1996-97, 1998-99, 2010-11, and 2011-12 reports a very low gain of area under rubber plantation which is below 150 ha. These suggest that there is no uniformity over the years in the annual gain of area under rubber plantation developed by the TRPCL.

The trend line (taking 3 yearly moving average) of the annual gain in area under rubber plantation developed by the TRPCL is shown in figure 4.7. The trend line suggest a significant increase in the annual gain of area under rubber plantation developed by the TRPCL especially in its initial years reaching its peak around 1988-89. After 1989 there has been a substantial decline in the acquisition of area under rubber plantation by this agency and the annual gain of area under rubber plantation is marginal till the year 1991-92. After 1992, the annual gain of area under rubber plantation picked up till 1994-95. From the year 1995-96 the annual gain of area under rubber plantation is suggestive of fluctuating trend till 1997-98. After 1998, the annual gain of area under rubber plantation started increasing till 2002-03. However, after 2004, the annual gain of area under rubber plantation developed by the TRPCL started increasing again reaching a second peak around 2007-08. After 2008, the gain in area under rubber plantation maintained by this agency declined significantly till 2011-12. However, from 2012-13 onward the annual gain of area under rubber plantation again suggest an increasing trend line (appendix 7). The trend line suggests that there are two peaks in the annual gain of area under rubber plantation developed by the TRPCL where the first peak is noticed during 1988-89 and the second peak is noticed during 2007-08.

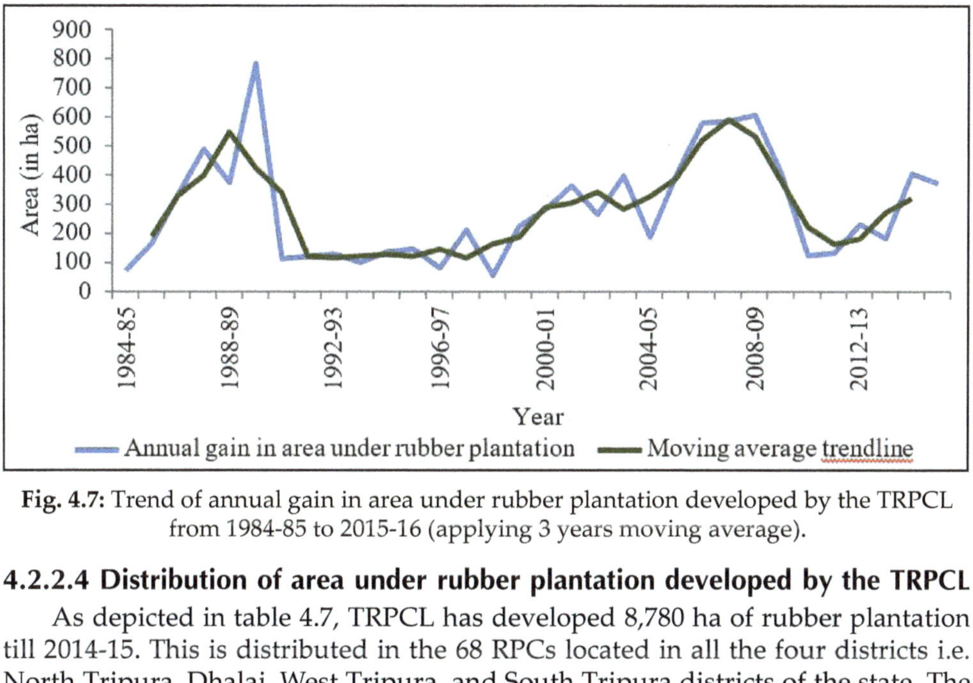

Fig. 4.7: Trend of annual gain in area under rubber plantation developed by the TRPCL from 1984-85 to 2015-16 (applying 3 years moving average).

4.2.2.4 Distribution of area under rubber plantation developed by the TRPCL

As depicted in table 4.7, TRPCL has developed 8,780 ha of rubber plantation till 2014-15. This is distributed in the 68 RPCs located in all the four districts i.e. North Tripura, Dhalai, West Tripura, and South Tripura districts of the state. The maximum area under rubber plantation developed by this agency is found in the West Tripura district (41% of the total). Here there are 19 RPCs. This is followed by South Tripura district which accounts for 23.48 percent of the total area under rubber plantation developed by this agency. Here there are 17 RPCs. About 18 percent of the total area under rubber plantation developed by this agency is found in North Tripura district where 16 RPCs are located. In Dhalai district about 17 percent of the area under rubber plantation developed by this agency is located where 16 RPCs are functioning.

Table 4.7: District wise distribution of area under rubber plantation developed by the TRPCL (2014-15)

District	Area under rubber plantation (in ha)	Share in %	RPC (in no)	Share in %
North Tripura	1,582	18.02	16	23.53
Dhalai	1,526	17.38	16	23.53
West Tripura	3,610	41.12	19	27.94
South Tripura	2,062	23.48	17	25.00
Total	8,780	100	68	100

Source: TRPCL, 2016.

In case of district wise share of area under rubber plantation developed by the TRPCL to the district wise total area under rubber plantation of the state during 2014- 15 is depicted in table 4.8.

Table 4.8: District wise share of area under rubber plantation developed by the TRPCL to the district wise total area under rubber plantation of the state (2014-15)

District	Area under rubber plantation (in ha) in the state	Area rubber plantation (in ha) of TRPCL	Share of area under rubber plantation (in %) of the TRPCL
North Tripura	9,365	1,582	16.89
Dhalai	3,874	1,526	39.39
West Tripura	35,317	3,610	10.22
South Tripura	21,740	2,062	9.48

Source: Rubber Board, 2016; TRPCL, 2016 (share of area under rubber plantation of the TRPCL computed by the researcher).

It is observed that TRPCL has maximum contribution of area under rubber plantation in Dhalai district with 39.39 percent followed by North Tripura, West Tripura and South Tripura districts with 16.89 percent, 10.22 percent and 9.48 percent respectively. It is noticed that although the TRPCL has less area under rubber plantation in Dhalai district than North Tripura, West Tripura and South Tripura districts but highest contribution of area under rubber plantation is found in Dhalai district. The reason may be due to less contribution of area under rubber plantation by other agencies and private owners in Dhalai district and more in North Tripura, West Tripura and South Tripura districts.

4.2.2.4.1 District wise concentration of area under rubber plantation developed by the TRPCL

In Tripura, TRPCL has developed rubber plantation in all the four districts of the state (table 4.7 and 4.8). In order to measure the district wise concentration of area under rubber plantation developed by the TRPCL, concentration index (CI) score has been computed which is further mapped as shown in figure 4.8. The CI score value of the different districts ranges from 1.44 to 0.76. The concentration of area under rubber plantation has been categorised into very high concentration, high concentration, moderate concentration and low concentration.

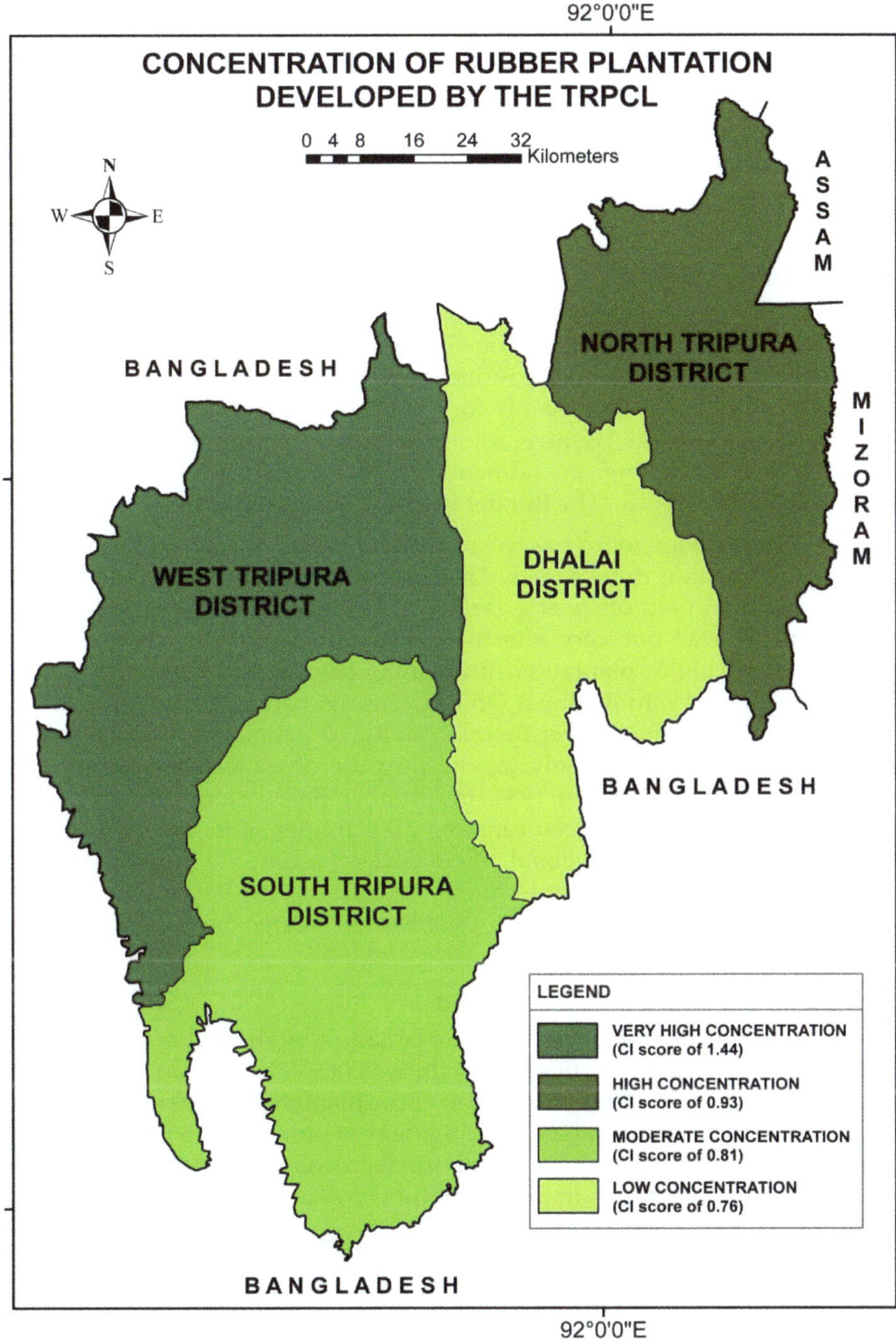

Fig. 4.8: District wise concentration of area under rubber plantation developed by the TRPCL (2014 -15).

As shown in figure 4.8, it is found that West Tripura district has the highest concentration of area under rubber plantation developed by the TRPCL as the CI score is 1.44, this is followed by North Tripura district where the area under rubber plantation developed by this agency is moderately concentrated having a CI value of 0.93. South Tripura district has low concentration of area under rubber plantation developed by this agency with a CI value of 0.81 while Dhalai district has the least concentration of area under rubber plantation having a CI value of 0.76.

4.3.3 The Rubber Board

The next important agency which looks after the rubber plantation programme of the state is the Rubber Board. The Government of India constituted "Indian Rubber Board" under the Rubber (Production and Marketing) Act 1947 which came into force on 18th April 1947, mainly for the overall promotion and development of rubber industry in the country, under its guidance and control. The Rubber Production and Marketing amendment Act of 1st August 1954), renamed the "Indian Rubber Board" as "The Rubber Board" (Rubber Board).

In 1963, the Rubber Board has technically supported the Forest Department of Tripura for cultivating rubber trees. The success of rubber trees made the Board set up its one man field office at Agartala in 1967 for undertaking development activities, but it was not very effective. However, seeing the potential in the development of rubber plantation (Bhowmik, 2006) in 1979 the one-man field office was upgraded into Regional Office (Viswanathan and Bhowmik, 2014). In the same year (1979), the Rubber Board established a regional research station at Taranagar covering 85 ha mainly for the purpose of achieving location specific R&D (Rubber Board, 2016). In 1984, the Rubber Board setup it zonal office at Agartala for monitoring and coordinating all activities of its offices in the state. The Rubber Board has four regional offices under the zone viz. Agartala, Udaipur, Dharmanagar and Ambassa. Under these regional offices, eleven field stations at various strategic locations have been established to cater the need of the rubber growers (Rubber Board, 2016).

4.3.3.1 Functioning of the Rubber Board

The expansion of area under rubber plantation by the Rubber Board started mainly through private initiatives (Bhowmik, 2006; Viswanathan and Bhowmik, 2014). However, after the implementation of the plantation development schemes, the Rubber Board started initiating cultivation of rubber trees out of the estate sector to the rural mass (very small to marginal growers) so as to bring the balance area under rubber plantation in the state (Rubber Board, 2016).

The Rubber Board supports most of the small and marginal farmers for setting up of rubber trees by equipping them with not only technical support but also providing them with financial assistance (Rubber Board, 2016) which are as follow:-

i) Provide assistance of Rs 30,000 per ha for new planting and replanting of rubber trees,

ii) Provide assistance of Rs 5,000 to 12,000 per ha for setting fencing/boundary,

iii) Provide assistance of Rs 13,500 per ha for the purpose of rejuvenation and revitalization of sick rubber trees,

iv) Provide assistance of Rs 20,000 per ha for restocking of the damaged rubber trees, centre,

v) Provide assistance of Rs 70,000 for construction of community processing

vi) Provide productivity enhancement scheme like supply of chemical fertilizers and rain-guarding materials at subsidized rates,

vii) Generate and supply quality planting materials, and

viii)Provide welfares schemes for the labourers.

Apart from the these supportive measures, the Rubber Board also has established Regional Rubber Training Centre in 2000 at Agartala mainly for promoting knowledge to the stakeholders of rubber, training programmes on estate management, crop harvesting, apiculture etc in order to familiarise the culture of rubber plantation to the rubber growers. The growers are also offered practical training programme on tapping and processing sponsored by Rubber Mission and different state government undertakings on a regular basis in village level meeting and promotional campaigns so as to improve the proficiency and adoption of innovation technology. The Rubber Board has also introduced Rubber Producers Societies (PRSs) to encourage the growers for self-reliance in maintaining their rubber plantation, tapping and processing, marketing, ensuring better price for the latex, rubber sheets, and rubber scraps through collective bargain, to monitor the market trends and also update market intelligence. At present, there are 156 RPSs, where the rubber growing community of the state are receiving commendable service (Rubber Board, 2016).

The Rubber Board has also initiated various schemes for the welfare of tappers and plantation workers like education stipend scheme, housing subsidy scheme, medical attendance scheme, and group insurance cum deposit scheme (Rubber Board, 2016).

The Rubber Board also offers 50 percent subsidy limited to Rs 700,000 to the RPSs for setting up of community processing centres so to overcome financial difficult of the small growers during processing and maintain the quality of rubber produce. At present, ninety-three community processing centres has been setup (50 groups and 43 blocks). Apart from this subsidy, the Rubber Board also offers financial assistance to individual growers for purchasing sheeting rollers at Rs 10,000 per unit and for construction of smoke houses at Rs 20,000 per unit (Rubber Board, 2016).

The Rubber Board has also setup its own nurseries so to produce and distribute quality planting and genuine budded stamps and bud wood materials. The Rubber Board has also established regional research station at Agartala in order to take up investigations on various aspects of rubber cultivation like location specific rubber clones suitable for the state and to develop appropriate agro-management practices (Rubber Board, 2016).

The strong network of dealers throughout the state coupled with the bargaining power of the RPSs and market intervention of the Rubber Board through its companies i.e. M/S Manimalayar Rubber (p) Ltd. has made rubber plantation the only crop which enjoys a farm gate price above 95 percent. The Rubber Board looks into the strengthening of the RPS by acting as an apex body of RPSs and assisting marketing of crop, ensuring better price, the company is also involved in distribution of almost all estate inputs required by growers at reasonable price in the state (Rubber Board, 2016).

4.3.3.2 Rehabilitation scheme of the Rubber Board

Apart from providing financial assistance and subsidy to the small and marginal farmers for cultivation of rubber trees, the Rubber Board has also implemented Block Plantation Scheme (BPS)/Block Plantation Project (BPP) since 1992 in order to rehabilitate the *jhumia* families so to uplift their socio-economic condition and also to afforest the denuded forestland through rubber plantation (Rubber Board, 2016).

Some of the silent features of the Rubber Board and structure of the project (Government of India (GOI), 2003) are-

i) Rubber Board appoints the Block Plantation Officer (the focal point in the BPP) to ensure all inputs required for rubber plantation are readily available and facilitates coordination with local authorities and the state government,

ii) Several state government schemes like immunisation drives, health camps, and animal husbandry camps are organised for the families in the project, and

iii) About 80 percent of the costs in the project are shared by the Rubber Board and the government of Tripura whereas about 20 percent are contributed by the beneficiaries as labour in the project.

4.3.3.2.1 Functioning of the project of the Rubber Board

For rehabilitation programme under the project, *firstly,* a compact land (block) owned by the ST/SC households are identified and then the Rubber Board initiates the planting rubber trees by engaging the beneficiary families as wage labour. Some of the important criteria noted while raising rubber plantation under the project (Rubber Board, 1992, 1999, 2005, cited by Viswanathan and Bhowmik, 2014) are-

i) Minimum extent of land available in the colony should be around 50 ha,

ii) The land should be in contiguous plots,

iii) ST beneficiaries who own land should hand over the land to the Rubber Board for a period of seven years,

iv) Assurance should be undertaken by the beneficiaries for the engagement of family labour during the immature phase of rubber plantation,

v) The beneficiaries should have clear title to the land,

vi) The extent of land should range from about 1 ha to 2 ha,

vii) Formation of RPS should be initiated after maturity of rubber plantation,

viii)The plantation is to be retained by the Rubber Board for two more years after opening the trees for stabilizing harvesting, processing and marketing procedures before handing over the plots back to the beneficiaries, and

ix) The financial expenditure will be borne by the state government and the Rubber Board while beneficiaries' contributions should be in the form of labour.

Apart from the development of rubber plantation through the project, the Rubber Board also implements various services like village roads, balwadis, health care centres, and drinking water schemes. The RPS and women thrift groups (WTGs) monitor the schemes for its proper functioning. Here, the RPS consists of male member of the beneficiaries and formulates many strategies like protection of the rubber plantation, takes up income-generation activities and acts as a forum for dissemination of knowledge and transfer of technology. The RPS also organises monthly meetings to discuss issues associated with rubber, initiates self-help groups (SHGs), engaged in rubber production including latex collection, processing, marketing and distributing of sales proceeds. The WTGs on the other hand plays an important role for women upliftment (GOI, 2003). The project also provides a platform for effective implementation of extension programmes, technology transfer and help in linking the local peasantry to main stream development process (Sharma, George K. and Dey, 2014). Consequently, the project has been considered as an important agency in paving the way for numerous tangible benefits including wage employment opportunities to the backward communities and providing sustainable means of livelihood (GOI, 2003).

The project implemented by the Rubber Board is considered to be a success. The success has resulted to positive participation of the local peasantry as the project provides different services on scientific plantation, followed by efficient processing and marketing of the produce. It also supplies the necessary inputs, equipment and services to the growers through the institutionalized set-up of RPS network (Sharma, George K. and Dey, 2014).

4.3.3.3 Rubber Board and the emerging pattern of increase of area under rubber plantation

The Rubber Board plays an important role in the expansion of rubber plantation in the state by providing financial assistance even in the rubber plantations which are not directly controlled by the Rubber Board. The rubber plantation developed for rehabilitation programme by the Rubber Board has been discussed as follows.

4.3.3.3.1 Share of area under rubber plantation developed by the Rubber Board to the total area under rubber plantation of the state

The share of area under rubber plantation developed by the Rubber Board to the total area under rubber plantation of the state during its inception (1992-93) was just 1.92 percent which has increased to 5.08 percent during 2015-16 (appendix 8). The average share of area under rubber plantation developed by the Rubber

Board to the total rubber plantation area in the state (1992-93 to 2015-16) is to the tune of 7.03 percent.

The share of area under rubber plantation developed by the Rubber Board to the total area under rubber plantation of the state is shown in figure 4.9. The figure suggests that from 1992-93 onwards the share of area under rubber plantation developed by Rubber Board has significantly increased till the year 1998-99. After 1999, the share of area under rubber plantation developed by the Board suggests steady increase reaching its highest share in the year 2003-04. After 2004, the share of area under rubber plantation developed by the Rubber Board to the total area under rubber plantation of the state started declining continuously till the year 2015-16. The share of area under rubber plantation of the Rubber Board to the total area under rubber plantation of the state is declining in the recent years and this is due to the entry of private sector and individual owners who have come up bringing more area under rubber plantation in the state. It is noticed that the role of Rubber Board in the development of area under rubber plantation in the state of Tripura is not much when compared to the other two agencies viz. TFDPCL and TRPCL.

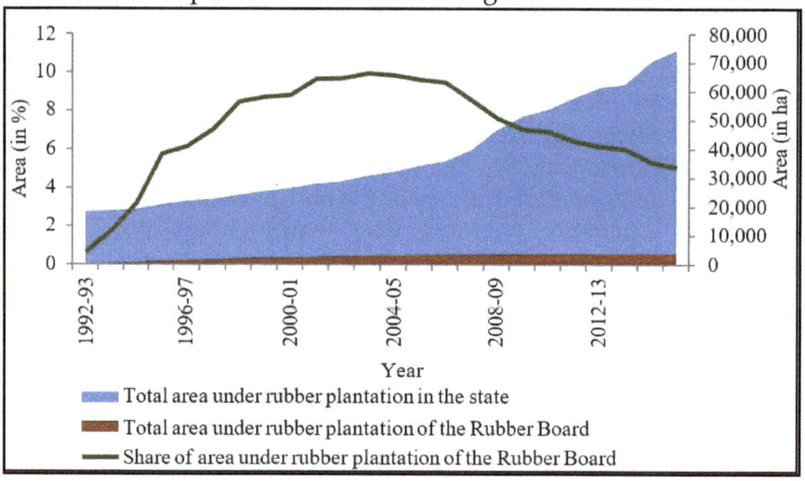

Fig. 4.9: Share of area under rubber plantation developed by the Rubber Board to the total area under rubber plantation of the state (1992-93 to 2015-16).

4.3.3.3.2 Role of Rubber Board in the expansion of area under rubber plantation

In the initial year (1992-93) the Rubber Board developed only 122 ha of rubber plantations which during 2015-16 has increased to 3,775 ha (appendix 9). From 1992-93 to 2015-16, the Rubber Board has developed 3,775 ha of area under rubber plantation that suggests the overall growth rate of area under rubber plantation as high as 2,2994.26 percent within a span of about twenty seven years. Figure 4.10 clearly indicates that there has been a significant expansion of area under rubber plantation developed by the Rubber Board from 1992-93 till 1998-99 however after 1999, area under rubber plantation developed by the Rubber Board is expanding at a slower pace.

The continuous increased in area under rubber plantation developed by the Rubber Board is attributed to the introduced of BPS/TBPP since 1992 in collaboration with the Rubber Board (pioneer in implementing the Project) and the Tribal Welfare Department, government of Tripura with sponsorship from the World Bank basically to promote economic settlement of the ST *jhumia* families through rubber plantation (Bhowmik, 2006; Viswanathan and Bhowmik, 2014). Deficiency of funding for rubber plantation based BPS/TBPP may be the reason for no extension in rubber plantation by the Rubber Board since 2010-11.

To show the trend in the expansion of area under rubber plantation developed by the Board, the annual growth of area under rubber plantation has been computed which is depicted in figure 4.10. The figure clearly suggests that area under rubber plantation developed by the Rubber Board is increasing continuously but at a declining rate. The highest annual growth rate of area under rubber plantation developed by the Rubber Board is noticed in the year 1993-94 with an annual growth rate of 176.23 percent while the lowest annual growth rate is observed in the year 2012-13 with an annual growth rate of as low as 1.34 percent (appendix 9). The average annual growth rate of area under rubber plantation developed by the Rubber Board is 20.72 percent.

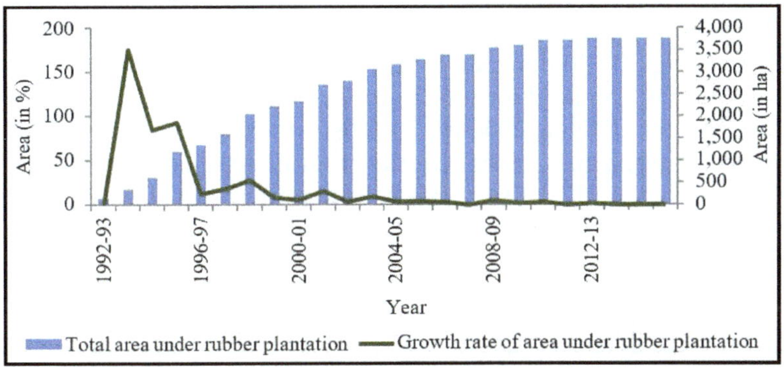

Fig. 4.10: Growth rate of area under rubber plantation developed by the Rubber Board (1992-93 to 2015-16).

4.3.3.3.3 Role of Rubber Board in the annual gain of area under rubber plantation

The highest annual gain of area under rubber plantation developed by the Rubber Board is observed in the year of 1995-96 where the annual gain was as high as 581 ha while the least annual gain of area under rubber plantation is noticed during 2012-13 with a gain of just 50 ha (appendix 9). The high annual gain in area under rubber plantation developed by the Rubber Board during the initial phase may be due to available land for rubber plantation and target initiated for rehabilitation programme through rubber plantation.

The average annual gain of area under rubber plantation from 1992-93 to 2015-16 developed by the Rubber Board is to the tune of 157.29 ha. The standard deviation of annual gain in area under rubber plantation developed by the Rubber Board from 1992-93 to 2015-16 is 148.55 and the coefficient of variation is to the tune of 94.44. This suggests that there is a large variation in the annual gain of area under rubber plantation developed by the Rubber Board from 1992-93 to 2015-16.

The annual gain in area under rubber plantation developed by the Rubber Board from 1992-93 to 2015-16 range from 56 ha (lowest) to 581 ha (highest). Consequently, the nature of gain in area under rubber plantation developed by the Rubber Board has been grouped into four categories as shown in table 4.9.

Table 4.9: Nature of gain in area under rubber plantation and number of years (developed by the Rubber Board from 1992-93 to 2015-16)

Nature of gain	Area under rubber plantation (ha)	No. of year
Very high gain	Above 300 ha	3
High gain	201 – 300 ha	4
Moderate gain	101 – 200 ha	9
Low gain	Below 100 ha	3

Source: Computed by the researcher.

It is observed that very high gain in area under rubber plantations are observed in three numbers of year viz. 1995-96, 1998-99 and 2001-02 where the annual gain in area under rubber plantation is above 300 ha. Four numbers of year viz. 1993-94, 1994-95, 1997-98 and 2003-04 suggests a high gain in area under rubber plantation ranging 201-300 ha per year. A moderate annual gain in area under rubber plantation is observed for nine numbers of year viz. 1992-93, 1996-97, 1999-00, 2000-01, 2004-05, 2005-06, 2006-07, 2008-09, and 2010-11 where the annual gain in area ranges between 101-200 ha. On the other hand, three numbers of years i.e. 2002-03, 2009-10 and 2012-13 suggests a low annual gain in area under rubber plantation where the annual gain in area under rubber plantations is below 100 ha per year.

The trend line (taking 3 yearly moving average) of the annual gain in area under rubber plantation developed by the Rubber Board is shown in figure 4.11.

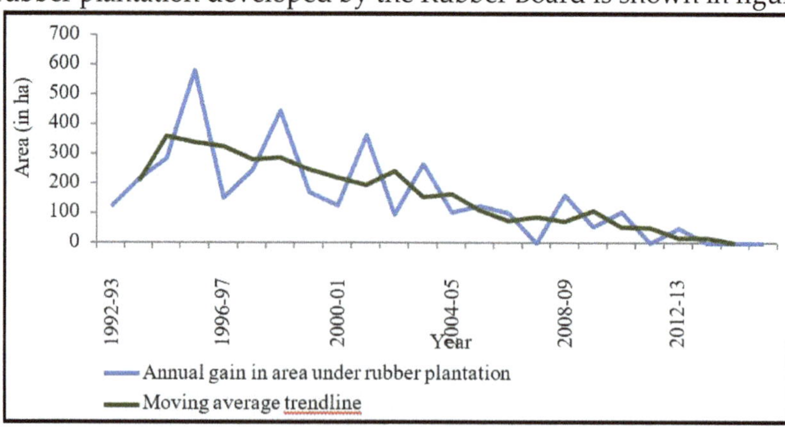

Fig. 4.11: Trend of annual gain in area under rubber plantation developed by the Rubber Board from 1992-93 to 2015-16 (applying 3 years moving average).

The trend line suggests that there is a significant increase in area under rubber plantation developed by the Rubber Board which reaches its peak in the year 1994-95. After 1995, the annual gain in area under rubber plantation developed by the Rubber Board shows a declining trend till 1997-98. From 1998-99 the annual gain in area under rubber plantation is suggestive of a fluctuating trend till 2015-16 where two peaks are observed in the year of 2002-03 and during 2009-10 (figure 4.11 and appendix 9).

4.3.3.4 Distribution of area under rubber plantation developed by the Rubber Board

Till 2014-15 the Rubber Board has developed 3,775 ha of rubber plantation which is distributed in 59 RPCs (table 4.10). The Rubber Board has developed rubber plantation in three districts i.e. North Tripura, West Tripura and South Tripura districts of the state. The Rubber Board has the highest area under rubber plantation (77.62 %) in West Tripura district which is distributed in 44 RPCs located here. West Tripura district is followed by South Tripura district where the Rubber Board has set up about 16.74 percent of rubber plantation distributed in 10 RPCs while the least area under rubber plantation developed by the Rubber Board is noticed in North Tripura district with just 5.64 percent of rubber plantation distributed in the 5 RPCs maintained by the Board. It is observed that the Rubber Board does not function in Dhalai district as it has no area under rubber plantation here.

Table 4.10: District wise area under rubber plantation developed by the Rubber Board (2014-15)

District	Area under rubber plantation (in ha)	Share in %	RPC (in no)	Share in %
North Tripura	213	5.64	5	8.47
Dhalai	----	-----	---	-----
West Tripura	2,930	77.62	44	74.58
South Tripura	632	16.74	10	16.95
Total	3,775	100	59	100

Source: Rubber Board, 2016.

In case of district wise share of area under rubber plantation developed by the Board to the district wise total area under rubber plantation of the state during 2014- 15 as depicted in table 4.11 indicates that Rubber Board has large contribution of area under rubber plantation in West Tripura district (8.30 %) followed by South Tripura and North Tripura districts with 2.91 percent and 2.27 percent respectively.

Table 4.11: District wise share of area under rubber plantation developed by the Rubber Board to the district wise total area under rubber plantation of the state (2014-15)

District	Area under rubber plantation (in ha) in the state	Area under rubber plantation (in ha) of the Rubber Board	Share of area under rubber plantation (in %) of the Rubber Board
North Tripura	9,365	213	2.27
Dhalai	3,874	-------	-----
West Tripura	35,317	2,930	8.30
South Tripura	21,740	632	2.91

Source: Rubber Board, 2016 (hare of area under rubber plantation of the Rubber Board computed by the researcher).

4.3.3.4.1. District wise concentration of area under rubber plantation developed by the Rubber Board

In Tripura, Rubber Board has developed rubber plantation in only three districts of the state (table 4.10 and 4.11). With the help of concentration index (CI), concentration of area under rubber plantation developed by the Rubber Board has been computed for the year of 2014-15, which has been further mapped as shown in figure 4.12. The CI score of the different districts ranges between 2.04 to 0.22. The concentration of area under rubber plantation has been categorized into high concentration, moderate concentration and low concentration.

Figure 4.12 suggests that West Tripura district has the highest concentration of area under rubber plantation developed by the Rubber Board as the CI score is 2.04. This is followed by South Tripura district where the area under rubber plantation is moderately concentrated with a CI score of 0.43 while North Tripura district has low concentration of area under rubber plantation with a CI value of 0.22.

4.4.4 Role of other agencies

Besides the TFDPCL, TRPCL and the Rubber Board other agencies which has also play a role in the expansion of area under rubber plantation in the state are the office of the TTAADC, office of the SDM/SDO and the Tribal Welfare Department which are also taking initiatives in the rehabilitation programme for the marginalized section of people through rubber plantation; however, their contribution is marginal when compared to the three main agencies viz. TFDPCL, TRPCL and the Rubber Board.

4.3 Comparison among the different agencies in the expansion of area under rubber plantation in Tripura

It is observed from table 4.12 that among the different agencies, TFDPCL has played a significant role in the expansion of area under rubber plantation as it has developed 11,624 ha of rubber plantation; TRPCL has developed 9,154 ha of rubber plantation while Rubber Board has contributed to the development of 3,775 ha of rubber plantation till 2015-16. However, others (private sector and individual owner) have developed about 49,782 ha of rubber plantation till 2015-16 which is very high as compared to the area under rubber plantation developed by three major agencies together. The increase in area under rubber plantation by the private sector and individual owner are mainly due to the increase in price of NR in the recent decade which attracts private sector and individual entrepreneurs to rubber plantation, consequently, resulting to expansion of area under rubber plantation in the state.

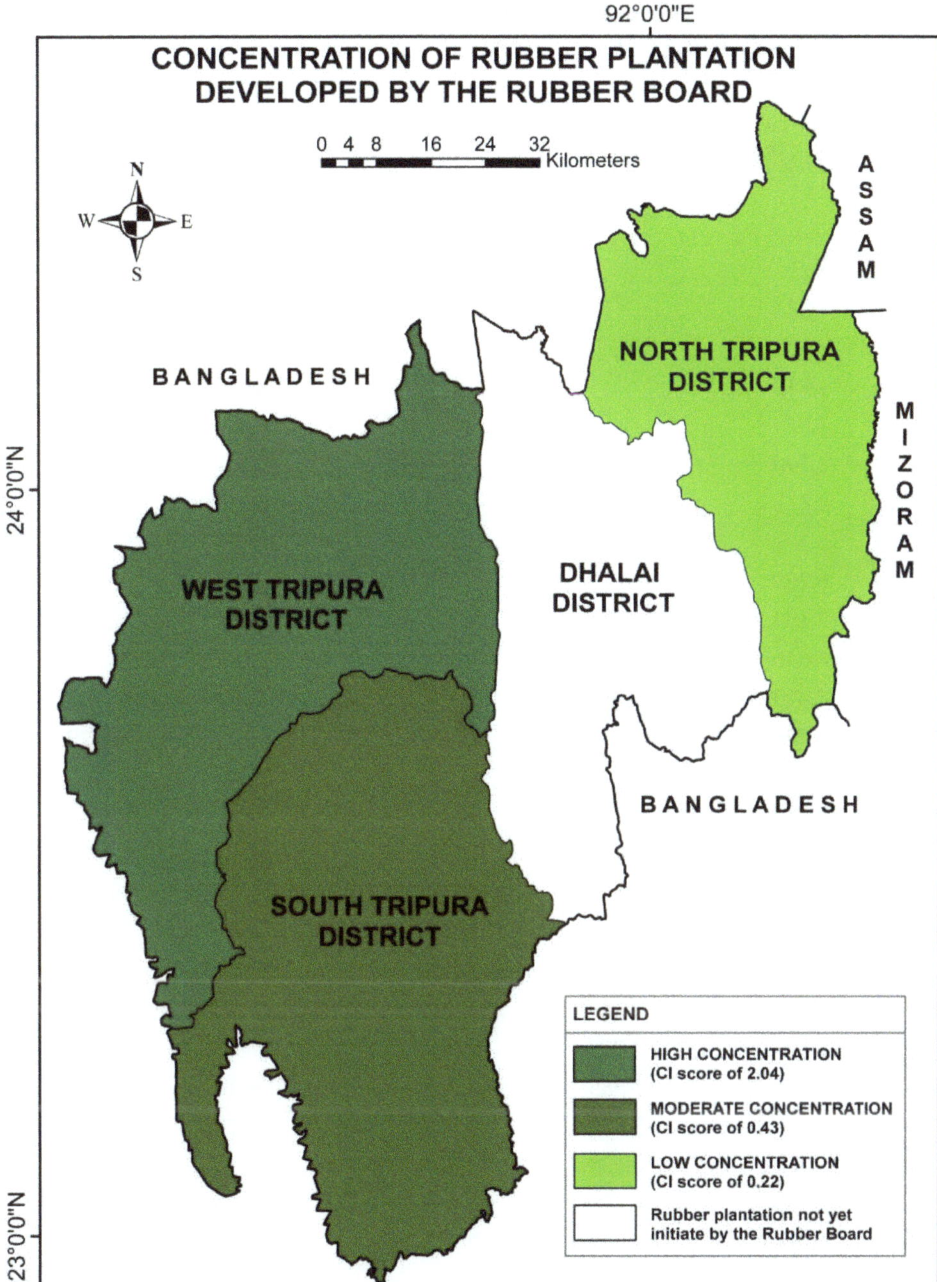

Fig. 4.12: District wise concentration of area under rubber plantation developed by the Rubber Board (2014-15).

Table 4.12: Role of different agencies in the expansion of area under rubber plantation in Tripura (1976-77 to 2015-16)

Year	Total area (in ha) of the TFDPCL	Total area (in ha) of the TRPCL	Total area (in ha) of the Rubber Board	Total area (in ha) of others*	Total area (in ha) in the state
1976-77	148			426	574
1977-78	451			427	878
1978-79	867			439	1,306
1979-80	1,534			455	1,989
1980-81	2,232			514	2,746
1981-82	2,948			642	3,590
1982-83	3,673			905	4,578
1983-84	4,190			1,148	5,338
1984-85	4,825	75		1,645	6,545
1985-86	5,361	245		2,136	7,742
1986-87	6,265	580		3,240	10,085
1987-88	7,169	1,072		4,429	12,670
1988-89	7,646	1,449		5,231	14,326
1989-90	8,620	2,236		5,764	16,620
1990-91	8,709	2,352		6,259	17,320
1991-92	8,840	2,476		6,544	17,860
1992-93	8,971	2,608	122	6,549	18,250
1993-94	9,059	2,711	337	6,563	18,670
1994-95	9,158	2,851	623	6,620	19,252
1995-96	9,390	3,000	1,204	7,167	20,761
1996-97	9,479	3,084	1,356	8,063	21,982
1997-98	9,537	3,299	1,602	8,144	22,582
1998-99	9,566	3,358	2,049	9,147	24,120
1999-00	9,614	3,585	2,220	9,961	25,380
2000-01	9,757	3,870	2,346	10,522	26,495
2001-02	9,787	4,237	2,710	11,213	27,947
2002-03	9,929	4,505	2,806	11,613	28,853
2003-04	10,093	4,905	3,073	12,699	30,770
2004-05	10,192	5,095	3,176	13,602	32,065
2005-06	10,214	5,496	3,301	15,178	34,189
2006-07	10,594	6,079	3,402	15,685	35,760
2007-08	10,877	6,668	3,402	18,723	39,670
2008-09	11,040	7,278	3,564	24,706	46,588

Year	Total area (in ha) of the TFDPCL	Total area (in ha) of the TRPCL	Total area (in ha) of the Rubber Board	Total area (in ha) of others*	Total area (in ha) in the state
2009-10	11,083	7,689	3,620	28,900	51,292
2010-11	11,134	7,816	3,725	30,994	53,669
2011-12	11,181	7,952	3,725	34,763	57,621
2012-13	11,201	8,186	3,775	38,069	61,231
2013-14	11,203	8,372	3,775	39,179	62,529
2014-15	11,211	8,780	3,775	46,530	70,296
2015-16	11,624	9,154	3,775	49,782	74,335

*Source: Rubber Board, 2016; TFDPCL, 2016; TRPCL, 2016; Centre for Development Studies. Note: *Others includes private sectors, individual owner and others agencies (like SDM and TTAADC but their contribution is marginal).*

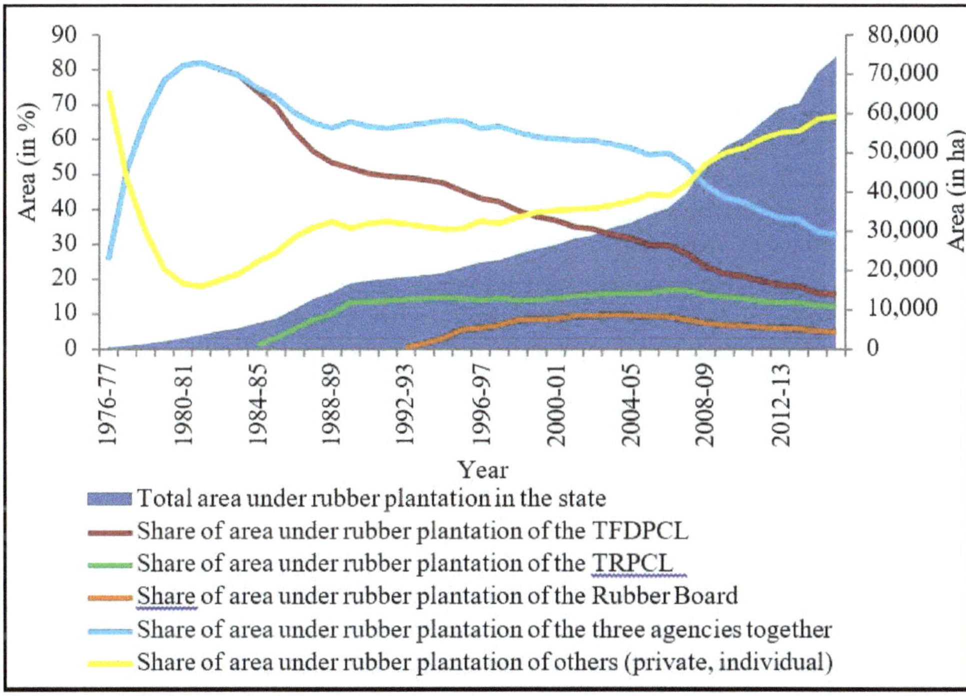

Fig. 4.13: Share of area under rubber plantation of the different agencies and others to the total area under rubber plantation of the state.

It is observed that the share of area under rubber plantation of the different agencies to the total area under rubber plantation of the state has been declining continuously in the recent years and the main reason are due to the increase in the share of area under rubber plantation by the private and individual owner to the total area under rubber plantation of the state which is clearly shown in figure 4.13.

It is found that although the annual gain of area under rubber plantation of the TFDPCL, TRPCL and the Rubber Board are declining but the total area under rubber plantation of the three agencies are increasing though at present only marginally. The expansion of area under rubber plantation of all these agencies are increasing but at a declining annual growth rate (table 4.12 and figure 4.13).

It is observed that though all the three agencies are found taking various initiatives in the rehabilitation programme of the marginalized section of people through rubber plantation; but TFDPCL is found to be more active in commercial gain rather than the rehabilitation of the marginalized as of the 11,211 ha of rubber plantation about 8,118 ha is for commercial purpose while only 3,093 ha is for resettlement purposes (TFDPCL, 2016a).

It is observed from table 4.13 that TRPCL has developed area under rubber plantation in all the four districts of the state i.e. North Tripura, Dhalai, West Tripura and South Tripura districts while both TFDPCL and the Rubber Board have developed rubber plantation only in three districts (North Tripura, West Tripura and South Tripura) and totally absent in Dhalai district.

District wise share of area under rubber plantation developed by the TFDPCL, TRPCL and Rubber Board to the district wise total area under rubber plantation of the state during 2014-15 as shown in table 4.13 indicates that TFDPCL has the highest share of area under rubber plantation in North Tripura district while least in West Trpura district. TRPCL has the highest share of area under rubber plantation in Dhalai district and notices the least share in South Tripura district. Rubber Board on the other hand has the least share in North Tripura district and the highest share in West Tripura district.

In case of district wise concentration of area under rubber plantation computed using concentration index as observed from figure 4. 4, 4.8 and 4.12 depicted that TFDPCL has high concentration of area under rubber plantation in South Tripura district. TRPCL has the high concentration of area under rubber plantation in West Tripura district while Rubber Board has high concentration of area under rubber plantation in West Tripura district. This suggest that both TRPCL and Rubber Board has high concentration of area under rubber plantation in West Tripura while TFDPCL in South Tripura districts.

4.4 Conclusion

It can be concluded that the two agencies, namely- TFDPCL and TRPCL are playing a significant role in initiating promotion and development of rubber plantation in the state of Tripura. The commercial gain associated with rubber plantation encourages private sector and individual growers to take up rubber plantation; as a result of which expansion of rubber plantation has gained momentum within the state.

The role of the state government, the different agencies as well as the stakeholders forming the local participation suggests that rehabilitation through rubber plantation programme has been a success throughout the state. The empowerment of the rubber growers through forming societies of their own has been the most important step towards empowerment. These societies hold regular meetings at certain intervals to discuss issues regarding plantation, nurturing of the estates, tapping, processing and other challenges resulting to the development of skill thereby human capital. This has minimized the role of government officials who at present usually facilities- to provide the technical know-how through training and workshops and assist in marketing. Even the cash subsidies are given directly to the societies without the involvement of any intermediaries. The entire process has resulted in tremendous confidence-building among the beneficiaries. Needless to say, the poor *jhumias* who provide the main recruitment base for the underground militant organizations are no longer turning to the underground except for those who still do not have any real livelihood option (Talukdar, 2007, cited by Bhaumik, 2015).

TFDPCL has not only checked degraded forestlands but also soil erosion, bio- diversity conservation as well as wildlife survival. Besides commercial rubber plantation, TFDPCL has also resettled poor families implemented under different schemes and projects of the state government through rubber plantation in order to improve their socio-economic status such as incomes, better housing, food, health and education (TFDPCL, 2016a).

Table 4.13: District wise share of area under rubber plantation developed by the TFDPCL, TRPCL and Rubber Board to the district wise total area under rubber plantation of the state (2014-15)

District	* Area under rubber plantation (in ha) in the state	** Area under rubber plantation (in ha) of TFDPCL	Share of area under rubber plantation (in %) of TFDPCL	*** Area rubber plantation (in ha) of TRPCL	Share of area under rubber plantation (in %) of the TRPCL	**** Area under rubber plantation (in ha) of the Rubber Board	Share of area under rubber plantation (in %) of the Rubber Board
North Tripura	9,365	2,762	29.49	1,582	16.89	213	2.27
Dhalai	3,874	-------	-------	1,526	39.39	-------	-----
West Tripura	35,317	3,462	9.80	3,610	10.22	2,930	8.30
South Tripura	21,740	4,987	22.94	2,062	9.48	632	2.91

*Source: * Rubber Board, 2016; ** TFDPCL, 2016; *** TRPCL, 2016; **** Rubber Board, 2016 (share of area under rubber plantation of TFDPCL, TRPCL and Rubber Board computed by the researcher)*

TRPCL activities such as meeting the beneficiaries during the initial plantation, granting permission to practice *jhuming* within plantation plots for an initial two years with additional income support through wages for weeding, providing training in rubber plantation techniques, involving the *jhumias* in decision making through their participation in the Beneficiaries Committees, granting ownership rights over rubber plantation land, and providing community facilities inside the plantation centres, have enhanced the economic and cultural protection of the *jhumias*, and helped the gradual transformation of their economy from *jhum* cultivation to settled cultivation (Gupta, 2000).

Kumar (2014) states that it is worth appreciating the service provided by the Rubber Board (Ministry of Commerce and Industry, Government of India) in providing need based financial and techno-managerial skills to these agencies, without which the success of this programme would not have been realized. Beside these three agencies the roles of NRETC has helped in facilitating communication between the Rubber Board and various stakeholders of rubber plantation industry. The Joint Rubber Production Commissioner, Agartala represents Rubber Board in the development committees and Directors of the Board of state government agencies like TFDPCL, TRPCL, Rubber Park etc all work together for the proper implementation of this programme (Rubber Board, 2016).

The success of schemes such as Block Plantation, RPDS, multi-pronged extension scheme and formation of RPS implemented by different agencies have not only helped in development and expansion of area under rubber plantation in the state but also brought prosperity among the poor villagers through sustainable flow of income, which have helped in containing insurgency problem through the rehabilitation process, afforestation of the degraded forests, generated employment opportunities, reduced the practices of *jhum*, improved the socio-economic conditions of the rehabilitated poor beneficiaries and contributed to the state revenue. Consequently, the next chapter highlights on the socio-economic conditions of the marginalized section of people and the changing livelihood options provided by rubber plantation in the state.

References

Bhowmik, I. 2006. A status on Rubber Plantations in Tripura, *Tripura Rubber Mission- Technical Bulletin*, Tripura, pp. 1-21.

Joseph, J., George K, T. and Dey, S.K. 2010. Report on the socio-economic impact of Natural Rubber cultivation under the block planting scheme in Tripura, *MPRA Paper No. 52370*, pp. 1-39.

Sharma, G., Joseph, J., George K, T. and Dey, S.K. 2011. Impact of Mahatma Gandhi National Rural Employment Guarantee Act on Rubber Block Plantation Scheme in Tripura, *Agriculture Economic Research Review Vol. 24*: 525-530.

Sharma, G., George K., T. and Dey, S.K. 2014. Natural Rubber Sector in Tripura: Role of Institutional Innovations, *Agricultural Economics Research Review, Vol. 27*, pp 17-24.

Das, R. 2014. Rubber Plantation: Tripura Scenario, in K. Shib (ed.) *Rubber in Tripura (1963-2013)- Golden Jubilee Celebration, Souvenir,* Bina Printers, Arundhatinagar, Agartala, pp. 25-33.

TFDPCL. 2016a. *Plan for Responsible Rubber wood and Bamboo Plantations Management (2013-14 to 2017-18)*, Agartala, Tripura, Prepared in consultation with Indian Institute of Forest Management, NEHRU NAGAR, BHOPAL, pp. 1-201.

DESPD. 2016. *Economic Review of Tripura (2015-16), 17th issue,* Government of Tripura, Agartala, Tripura, pp. 3-151.

Bhowmik, I. 2006. A status on Rubber Plantations in Tripura, *Tripura Rubber Mission- Technical Bulletin*, Tripura, pp. 1-21.

Viswanathan, P.K. and Bhowmik, I. 2014. Compatibility of Institutional Architecture for Rubber Plantation Development in North East India from the Comparative Perspective of Kerala, *NRPPD Discussion Paper No. 38,* Centre For Development Studies, Thiruvananthapuram, pp. 1-71.

TFDPCL. 2016a. *Plan for Responsible Rubber wood and Bamboo Plantations Management (2013-14 to 2017-18),* Agartala, Tripura, Prepared in consultation with Indian Institute of Forest Management, NEHRU NAGAR, BHOPAL, pp. 1-201.

Ibid, pp. 1-201.

Sinha, A. K. 2012. Rubber plantation in Northeast India, hopes vs concerns.

The Tripura Foundation. Retrieved from http://thetripurafoundation.org/art-rubber-north-east-hopes-vs-concern. Accessed on April 2015.

TFDPCL. 2016a. *Plan for Responsible Rubber wood and Bamboo Plantations Management (2013-14 to 2017-18),* Agartala, Tripura, Prepared in consultation with Indian Institute of Forest Management, NEHRU NAGAR, BHOPAL, pp, 1-201.

Das, R. 2014. Rubber Plantation: Tripura Scenario, in K. Shib (ed.) *Rubber in Tripura (1963-2013)- Golden Jubilee Celebration, Souvenir,* Bina Printers, Arundhatinagar, Agartala, pp. 25-33.

TRPCL. 2012. *Brief Note on Tripura Rehabilitation Plantation Corporation Ltd.*

Diary 2012, Agartala, Tripura, pp. 1-12.

TRPCL. 2016. *Documents on Tripura Rehabilitation Plantation Corporation Limited,* Agartala, Tripura.

Viswanathan, P.K. and Bhowmik, I. 2014. Compatibility of Institutional Architecture for Rubber Plantation Development in North East India from the Comparative Perspective of Kerala, *NRPPD Discussion Paper No. 38,* Centre For Development Studies, Thiruvananthapuram, pp. 1-71.

TFDPCL. 2016a. *Plan for Responsible Rubber wood and Bamboo Plantations Management (2013-14 to 2017-18),* Agartala, Tripura, Prepared in consultation with Indian Institute of Forest Management, NEHRU NAGAR, BHOPAL, pp. 1-201.

Ibid, pp. 1-201.

Ibid, pp. 1-201.

Ibid, pp. 1-201.

Centre for Development Studies. Area under rubber and average yield per hectare.

Retrieved from http://cds.edu/research/ru/nrppd/database-on-plantation/rubber/. Accessed 23 January 2016.

Rubber Board. 2016. *Role of Rubber Board in Growth of Rubber Plantation in NE,* The regional office, Agartala.

TFDPCL, 2016. *Documents on TFDPCL,* TFDPC Bhavan, Agartala, Tripura.

Ibid.

Das, R. 2014. Rubber Plantation: Tripura Scenario, in K. Shib (ed.) *Rubber in Tripura (1963-2013)- Golden Jubilee Celebration, Souvenir,* Bina Printers, Arundhatinagar, Agartala, pp. 25-33.

TFDPCL, 2016. *Documents on TFDPCL*, TFDPC Bhavan, Agartala, Tripura.

Rubber Board. 2016. *Role of Rubber Board in Growth of Rubber Plantation in NE*, The regional office, Agartala.

TFDPCL, 2016. *Documents on TFDPCL*, TFDPC Bhavan, Agartala, Tripura. TFDPCL. 2016a. *Plan for Responsible Rubber wood and Bamboo Plantations*

Management (2013-14 to 2017-18), Agartala, Tripura, Prepared in consultation with Indian Institute of Forest Management, NEHRU NAGAR, BHOPAL, pp. 1-201.

Debbarma, B.,2014. Tripura Rehabilitation Plantation Corporation Limiited, in K. Shib (ed.) *Rubber in Tripura (1963-2013)- Golden Jubilee Celebration, Souvenir*, Bina Printers, Arundhatinagar, Agartala, pp. 35-42.

TRPCL. 2012. *Brief Note on Tripura Rehabilitation Plantation Corporation Ltd.*

Diary 2012, Agartala, Tripura, pp. 1-12.

TRPCL. 2016. *Documents on Tripura Rehabilitation Plantation Corporation Limited*, Agartala, Tripura.

Viswanathan, P.K. and Bhowmik, I. 2014. Compatibility of Institutional Architecture for Rubber Plantation Development in North East India from the Comparative Perspective of Kerala, *NRPPD Discussion Paper No. 38*, Centre For Development Studies, Thiruvananthapuram, pp. 1-71.

Sinha, A. K. 2012. Rubber plantation in Northeast India, hopes vs concerns.

The Tripura Foundation. Retrieved from http://thetripurafoundation.org/art-rubber-north-east-hopes-vs-concern. Accessed on April 2015.

Debbarma, B. 2014. Tripura Rehabilitation Plantation Corporation Limiited, in K. Shib (ed.) *Rubber in Tripura (1963-2013)- Golden Jubilee Celebration, Souvenir*, Bina Printers, Arundhatinagar, Agartala, pp. 35-42.

Viswanathan, P.K. and Bhowmik, I. 2014. Compatibility of Institutional Architecture for Rubber Plantation Development in North East India from the Comparative Perspective of Kerala, *NRPPD Discussion Paper No. 38*, Centre For Development Studies, Thiruvananthapuram, pp. 1-71.

Debbarma, B. 2014. Tripura Rehabilitation Plantation Corporation Limiited, in K. Shib (ed.) *Rubber in Tripura (1963-2013)- Golden Jubilee Celebration, Souvenir*, Bina Printers, Arundhatinagar, Agartala, pp. 35-42.

TRPCL. 2012. *Brief Note on Tripura Rehabilitation Plantation Corporation Ltd.*

Diary 2012, Agartala, Tripura, pp. 1-12.

TRPCL. 2016. *Documents on Tripura Rehabilitation Plantation Corporation Limited*, Agartala, Tripura.

Debbarma, B. 2014. Tripura Rehabilitation Plantation Corporation Limiited, in K. Shib (ed.) *Rubber in Tripura (1963-2013)- Golden Jubilee Celebration, Souvenir*, Bina Printers, Arundhatinagar, Agartala, pp. 35-42.

TRPCL. 2016. *Documents on Tripura Rehabilitation Plantation Corporation Limited*, Agartala, Tripura.

TRPCL. 2012. *Brief Note on Tripura Rehabilitation Plantation Corporation Ltd.*

Diary 2012, Agartala, Tripura, pp. 1-12.

Debbarma, B. 2014. Tripura Rehabilitation Plantation Corporation Limiited, in K. Shib (ed.) *Rubber in Tripura (1963-2013)- Golden Jubilee Celebration, Souvenir*, Bina Printers, Arundhatinagar, Agartala, pp. 35-42.

Viswanathan, P.K. and Bhowmik, I. 2014. Compatibility of Institutional Architecture for Rubber Plantation Development in North East India from the Comparative Perspective of Kerala, *NRPPD Discussion Paper No. 38*, Centre For Development Studies, Thiruvananthapuram, pp. 1-71.

TRPCL. 2012. *Brief Note on Tripura Rehabilitation Plantation Corporation Ltd.*

Diary 2012, Agartala, Tripura, pp. 1-12.

Ibid, pp. 1-12.

Ibid, pp. 1-12.

Debbarma, B. 2014. Tripura Rehabilitation Plantation Corporation Limiited, in K. Shib (ed.) *Rubber in Tripura (1963-2013)- Golden Jubilee Celebration, Souvenir*, Bina Printers, Arundhatinagar, Agartala, pp. 35-42.

Ibid, pp. 35-42.

TRPCL. 2012. *Brief Note on Tripura Rehabilitation Plantation Corporation Ltd.*

Diary 2012, Agartala, Tripura, pp. 1-12.

Debbarma, B. 2014. Tripura Rehabilitation Plantation Corporation Limiited, in K. Shib (ed.) *Rubber in Tripura (1963-2013)- Golden Jubilee Celebration, Souvenir*, Bina Printers, Arundhatinagar, Agartala, pp. 35-42.

TRPCL. 2012. *Brief Note on Tripura Rehabilitation Plantation Corporation Ltd.*

Diary 2012, Agartala, Tripura, pp. 1-12.

Debbarma, B. 2014. Tripura Rehabilitation Plantation Corporation Limiited, in K. Shib (ed.) *Rubber in Tripura (1963-2013)- Golden Jubilee Celebration, Souvenir*, Bina Printers, Arundhatinagar, Agartala, pp. 35-42.

TRPCL. 2012. *Brief Note on Tripura Rehabilitation Plantation Corporation Ltd.*

Diary 2012, Agartala, Tripura, pp. 1-12.

Ibid, pp. 1-12.

Debbarma, B. 2014. Tripura Rehabilitation Plantation Corporation Limiited, in K. Shib (ed.) *Rubber in Tripura (1963-2013)- Golden Jubilee Celebration, Souvenir*, Bina Printers, Arundhatinagar, Agartala, pp. 35-42.

TRPCL. 2012. *Brief Note on Tripura Rehabilitation Plantation Corporation Ltd.Diary 2012*, Agartala, Tripura, pp. 1-12.

Das, R. 2014. Rubber Plantation: Tripura Scenario, in K. Shib (ed.) *Rubber in Tripura (1963-2013)- Golden Jubilee Celebration, Souvenir*, Bina Printers, Arundhatinagar, Agartala, pp. 25-33.

Debbarma, B. 2014. Tripura Rehabilitation Plantation Corporation Limiited, in K. Shib (ed.) *Rubber in Tripura (1963-2013)- Golden Jubilee Celebration, Souvenir*, Bina Printers, Arundhatinagar, Agartala, pp. 35-42.

Centre for Development Studies. *Area under rubber and average yield per hectare*. Retrieved from http://cds.edu/research/ru/nrppd/database-on- plantation/ rubber/. Accessed 23 January 2016.

Rubber Board. 2016. *Role of Rubber Board in Growth of Rubber Plantation in NE*, The regional office, Agartala.

TFDPCL, 2016. *Documents on TFDPCL*, TFDPC Bhavan, Agartala, Tripura.

Ibid. Ibid. Ibid.

Rubber Board. 2016. *Role of Rubber Board in Growth of Rubber Plantation in NE*, The regional office, Agartala.

TFDPCL, 2016. *Documents on TFDPCL*, TFDPC Bhavan, Agartala, Tripura.

Rubber Board. *The Rubber Tree*. Retrieved from http://rubberboard.org.in/ rubbercultivation.asp. Accessed on April 2015.

Bhowmik, I. 2006. A status on Rubber Plantations in Tripura, *Tripura Rubber Mission- Technical Bulletin*, Tripura, pp. 1-21.

Viswanathan, P.K. and Bhowmik, I. 2014. Compatibility of Institutional Architecture for Rubber Plantation Development in North East India from the Comparative Perspective of Kerala, *NRPPD Discussion Paper No. 38*, Centre For Development Studies, Thiruvananthapuram, pp. 1-71.

Rubber Board. 2016. *Role of Rubber Board in Growth of Rubber Plantation in NE*, The regional office, Agartala.

Ibid.

Bhowmik, I. 2006. A status on Rubber Plantations in Tripura, *Tripura Rubber Mission- Technical Bulletin*, Tripura, pp. 1-21.

Viswanathan, P.K. and Bhowmik, I.,2014. Compatibility of Institutional Architecture for Rubber Plantation Development in North East India from the Comparative Perspective of Kerala, *NRPPD Discussion Paper No. 38*, Centre For Development Studies, Thiruvananthapuram, pp. 1-71.

Rubber Board. 2016. *Role of Rubber Board in Growth of Rubber Plantation in NE*, The regional office, Agartala.

Ibid. Ibid. Ibid. Ibid. Ibid. Ibid. Ibid.

Government of India. 2003. Successful Governance Initiatives and Best Practices: Experiences from Indian States, *Planning Commission (Government of India), Human Development Resource Centre, UNDP*, Academic Foundation, New Delhi, pp. 50-54.

Viswanathan, P.K. and Bhowmik, I. 2014. Compatibility of Institutional Architecture for Rubber Plantation Development in North East India from the Comparative Perspective of Kerala, *NRPPD Discussion Paper No. 38*, Centre For Development Studies, Thiruvananthapuram, pp. 1-71.

Government of India. 2003. Successful Governance Initiatives and Best Practices: Experiences from Indian States, *Planning Commission (Government of India), Human Development Resource Centre, UNDP*, Academic Foundation, New Delhi, pp. 50-54.

Sharma, G., George K., T. and Dey, S.K. 2014. Natural Rubber Sector in Tripura: Role of Institutional Innovations, *Agricultural Economics Research Review, Vol. 27*, pp 17-24.

Government of India. 2003. Successful Governance Initiatives and Best Practices: Experiences from Indian States, *Planning Commission (Government of India), Human Development Resource Centre, UNDP*, Academic Foundation, New Delhi, pp. 50-54.

Sharma, G., George K., T. and Dey, S.K. 2014. Natural Rubber Sector in Tripura: Role of Institutional Innovations, *Agricultural Economics Research Review, Vol. 27*, pp 17-24.

Centre for Development Studies. *Area under rubber and average yield per hectare.* Retrieved from http://cds.edu/research/ru/nrppd/database-on- plantation/ rubber/. Accessed on 23 January 2016.

Rubber Board. 2016. *Role of Rubber Board in Growth of Rubber Plantation in NE*, The regional office, Agartala.

Bhowmik, I. 2006. A status on Rubber Plantations in Tripura, *Tripura Rubber Mission- Technical Bulletin*, Tripura, pp. 1-21.

Viswanathan, P.K. and Bhowmik, I. 2014. Compatibility of Institutional Architecture for Rubber Plantation Development in North East India from the Comparative Perspective of Kerala, *NRPPD Discussion Paper No. 38*, Centre For Development Studies, Thiruvananthapuram, pp. 1-71.

Rubber Board. 2016. *Role of Rubber Board in Growth of Rubber Plantation in NE*, The regional office, Agartala.

Ibid. Ibid, Ibid.

TFDPCL, 2016. *Documents on TFDPCL*, TFDPC Bhavan, Agartala, Tripura. TRPCL. 2016. *Documents on Tripura Rehabilitation Plantation Corporation Limited*, Agartala, Tripura.

Centre for Development Studies. *Area under rubber and average yield per hectare.* Retrieved from http://cds.edu/research/ru/nrppd/database-on- plantation/ rubber/. Accessed on 23 January 2016.

TFDPCL. 2016a. *Plan for Responsible Rubber wood and Bamboo Plantations Management (2013-14 to 2017-18)*, Agartala, Tripura, Prepared in consultation with Indian Institute of Forest Management, NEHRU NAGAR, BHOPAL, pp. 1-201.

Rubber Board. 2016. *Role of Rubber Board in Growth of Rubber Plantation in NE*, The regional office, Agartala.

TFDPCL, 2016. *Documents on TFDPCL*, TFDPC Bhavan, Agartala, Tripura. TRPCL. 2016. *Documents on Tripura Rehabilitation Plantation Corporation Limited*, Agartala, Tripura.

Rubber Board. 2016. *Role of Rubber Board in Growth of Rubber Plantation in NE*, The regional office, Agartala.

Bhaumik, S. 2015. Tripura: Counter-insurgency and Governance, in R. Samaddar (ed.) *Government of Peace: Social Governance, Security and the Problematic of Peace*, Ashgate Publishing Ltd., Wey Court East, England, pp. 79-106.

TFDPCL. 2016a. *Plan for Responsible Rubber wood and Bamboo Plantations Management (2013-14 to 2017-18)*, Agartala, Tripura, Prepared in consultation with Indian Institute of Forest Management, NEHRU NAGAR, BHOPAL, pp. 1-201.

Gupta, A.K. 2000. Shifting Cultivation and Conservation of Biological Diversity in Tripura, Northeast India, *Human Ecology, Vol. 28, No. 4*, pp. 605-629.

Kumar, A.K.K. 2014. Rubber Development Version 2.0 for North Eastern Region with a reference to Tripura, in K. Shib (ed.) *Rubber in Tripura (1963-2013)- Golden Jubilee Celebration, Souvenir*, Bina Printers, Arundhatinagar, Agartala, pp. 13-18.

Rubber Board. 2016. *Role of Rubber Board in Growth of Rubber Plantation in NE*, The regional office, Agartala.

5

Impact of Rubber Plantation on the Socio-Economic Conditions and Livelihoods of the Surveyed Households

1.1 Introduction

As already mentioned rubber plantation was initially introduced in the state of Tripura as a soil conservation measure on the degraded forestland caused due to the practice of *jhum* cultivation and large scale illegal felling of forest trees, but later the state government (through different agencies), the Rubber Board and other state government departments have encouraged rubber plantation as a rehabilitation programme to check *jhum* cultivation and bring the *jhumias* and other landless families under the ambit of settled agriculture so to improve their socio-economic and livelihood conditions. In this connection, the primary data and information suggesting their socio-economic conditions and livelihood practices has been collected from the selected sample households consisting of two sets, one set representing the BHHs and the other consisting of households not having rubber plantations (NBHHs) but inhabiting the same space. This has been further discussed, compared and analysed in this chapter with a focus on not only assessing the role of rubber plantation in realizing the socio-economic development of the BHHs but also the livelihood practices and associated diversity.

The first section of this chapter looks into the RPC/village wise tribe composition and the livelihood assets/capitals of the BHHs in order to analyze whether the rehabilitation programme through rubber plantation scheme has created disparities in the socio-economic development among the ST population. The second section deals with the socio-economic characteristics of both sets of households (BHHs and NBHHs) and the third section deals with the analysis of livelihood diversity of the BHHs and NBHHs.

As already mentioned, with the help of sustainable livelihood approach (SLA) of the sustainable livelihood framework developed by the department for international development (DFID, 1999) an attempt has been made to assess the five essential livelihood assets/capitals viz. natural, physical, human, financial and social assets/capitals associated with livelihood studies by the help of suitable indicators. The indicators of the five assets/capitals has been analyzed using the range equalization technique developed by UNDP method of Human Development Index (Ghosh et al., 2006) in order to create indices for each indicator and to aggregate them into composite index score for each of the five assets/capitals a scale free aggregate index score ranging between 0-1 representing each indicator of the five livelihood assets/capitals has been computed. The composite index score of the five assets/capitals of the BHHs and NBHHs has been compared in order to analyze if rubber plantation has in reality enhanced the socio-economic conditions of the BHHs by bringing about a change in their access to resources, a pre-requisite for socio- economic development. The composite index score of the five assets/capitals of the BHHs and NBHHs has also been represented by the livelihood asset pentagon in order to analyze and compare the strength and weakness of the five assets/capitals of both the BHHs and NBHHs. Further the composite score value of each asset/capital of both the BHHs and NBHHs has been compared to the theoretical norm score in order to ascertain the level of sustainability of each livelihood asset/capital.

5.2 RPC/village wise tribe composition and the livelihood assets/capitals of the BHHs

5.2.1 Tribe composition of the BHHs

As the rehabilitation programme through rubber plantation scheme addresses mainly the marginalized communities, the RPC/village wise ST compositions of the BHHs have been looked into. As per census of India 2011, in Tripura there are nineteen ST communities which constitute 31.8 percent of the total population of the state. The major ST communities are Tripura/Tripuri/Tippera, Reang, Jamatia, Chakma, Halam and Mog which constitutes about 28.27 percent of the state's population (Census of India, 2011).

It is observed from table 5.1 that from the four RPCs/villages surveyed; households belonging to three tribes i.e. Debbarma, Reang and Chakma tribes have been rehabilitated through rubber plantation scheme managed by the TRPCL. Debbarma tribe are mainly concentrated in three surveyed RPCs/villages i.e. Dakshin Dhumachhara (75.56 %), Pathaliaghat and Dakshin Taidu with 100 percent each respectively. Reang tribe constitute 24.44 percent of the total households surveyed and they are mainly found in Dakshin Dhumachhara RPC/village while in Depachhara RPC/village most of the households belong to the Chakma tribe (100 %).

Table 5.1: RPC/village wise distribution of tribe of the BHHs

RPC/ village Tribe	Depachhara	Dakshin Dhumachhara	Pathaliaghat	Dakshin Taidu
	Share (in %) of tribe of the BHHs (n=54)	Share (in %) of tribe of the BHHs (n=45)	Share (in %) of tribe of the BHHs (n=56)	Share (in %) of tribe of the BHHs (n=55)

Debbarma	-----	75.56	100	100
Reang	-----	-------	----	----
Chakma	100	24.44	----	----

Source: Field survey, 2016.

As observed from table 5.1, household belonging to Debbarma tribe is found in two RPCs/villages, Chakma tribe in one RPC/village while in Dakshin Dhumachhara about 76 percent of the household belong to Debbarma tribe and only about 24 percent of the households belong to Reang tribe. Field data suggests that there is preponderance of Debbarma tribe who constitute the BHHs.

In order to assess if the rehabilitation programme through rubber plantation scheme has created disparities in the socio-economic development among the tribes in the four RPCs/villages, the five livelihood assets/capitals i.e. human, social, natural, physical and financial assets/capitals have been measured with suitable indicators. Further with the help of range equalization method (UNDP method of Human Development Index) (Ghosh et al., 2006) a scale free aggregate index score ranging between 0-1 representing each indicator of the five livelihood assets/capitals has been computed. Consequently, a composite index score of the five assets/capitals has been computed, measured and compared and also diagrammatically represented by livelihood asset pentagon. As the SLA seeks to gain an accurate and realistic understanding of people's strengths through assessing the five livelihood assets/capitals it becomes essential to define the five assets/capitals. The definitions of the five assets/capitals as per (DFID, 1999) are as follow:-

Human asset/capital- this refers to skills, knowledge, ability to labour and good health that together enables people to pursue different livelihood practices and achieve better life

Social asset/capital- this refers to the social resources upon which people draw upon in pursuit of their different livelihood objectives, this is developed through social networks and connections, membership of group and associations and relationship of trust, reciprocity and exchanges makes social asset.

Natural asset/capital- this refers to the natural resource stocks from which resource flows and services useful for livelihoods are derived.

Physical asset/capital- this comprises of the basic infrastructure an goods needed to support livelihoods and meet the basic needs to make people more productive in order to achieve a better socio-economic development.

Financial asset/capital- this refers to the financial resources that people use in order to achieve their livelihood objectives.

5.2.2 Comparative analysis of the five livelihood assets/capitals of the BHHs living in the four surveyed RPCs/ villages

Human asset/capital

Human asset/capital is considered as one of the most essential asset/capital in livelihood studies as it is the main asset/capital through which income is earned for a living. The indicators to measure human asset/capital are the share of working population, literacy rate, traditional skills known and modern skills known. Human asset/capital of the BHHs of the four RPCs/villages as depicted in figure 5.1 suggest that the BHHs from Pathaliaghat RPC/village is better off in human asset/capital than other RPCs/villages. The composite index score of the human asset/capital of the four RPCs/villages indicates that Pathaliaghat RPC/village has the highest composite index score (0.43) this is closely followed by Dakshin Dhumachhara RPC/village (0.38 score) and Dakshin Taidu RPC/village (0.37 score) while it is observed that Depachhara RPC/village has the least human asset/capital as the composite index score 0.29 (appendix 10). High literacy rate (83.90 %) and modern skills known (100%) by the members of the household of the Pathaliaghat RPC/village attribute for the high human asset/capital reported by the BHHs of the Pathaliaghat RPC/village. 100 percent of the BHHs from Pathaliaghat and Dakshin Taidu RPCs/village belong to the Debbarma tribe. The Debbarma tribe also constitute more than 75 percent of the BHHs in Dakshin Dhumachhara RPCs/village. Consequently field data suggests that the Debbarma tribe is better off than the Chakma tribe living in Depachhara RPC/village.

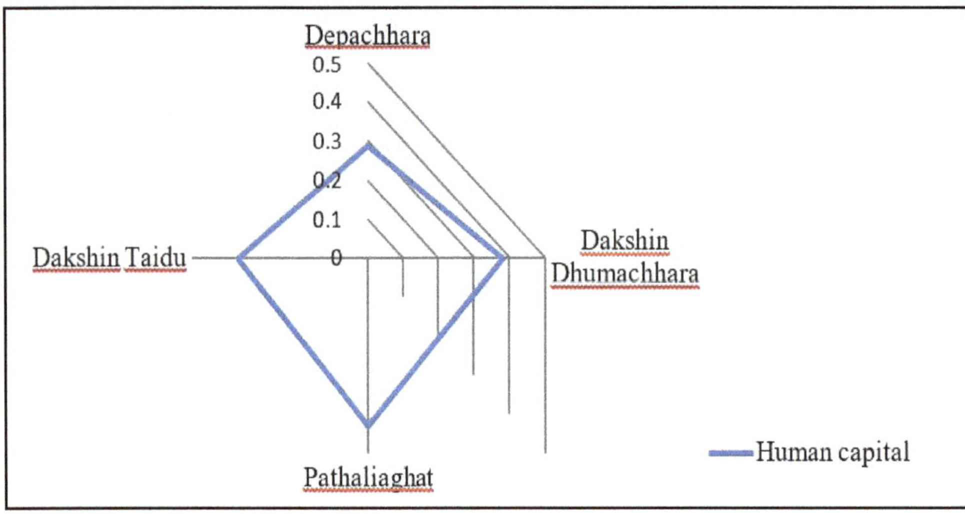

Fig. 5.1: Composite index score of human asset/capital of the BHHs of the RPCs.

Social asset/capital

Social asset/capital is important as it indicates the mutual trust and reciprocation among a group of people/community. The tribal society as such is usually rich in social capital as they are bound by their common kinship ties hence in this case the external indicators to social capital has been taken into consideration. The indicators to measure social asset/capital are good relationship with TRPCL officials of the RPC/village, training and capacity development activities, committee member in the locality, member in Self Help Group (SHG) and participation in programme or meeting organized by rubber growing agencies and others government and non-government organisation. In regard to social asset/capital of the BHHs in the four RPCs/villages, Dakshin Dhumachhara RPC/village has the highest score in social asset/capital in comparison to other RPCs/villages (figure 5.2). The composite index score of the social asset/capital of the four RPCs/villages as depicted in appendix 11, suggest that Dakshin Dhumachhara RPC/village has the highest composite index score (0.54), this is closely followed by Dakshin Taidu RPC/village (0.51 score) which is slightly higher than composite index score of Pathaliaghat RPC/village which has a score of 0.50 while Depachhara RPC/village has the least social asset/capital composite index score (0.47). BHHs from Dakshin Dhumachhara RPC/village reports higher participation in programme or meeting organised by rubber growing agencies and others government and non-government organisation (97.78 %) this has strengthened their social asset/capital. Social asset/capital of the four RPCs/villages as depicted in figure 5.2 suggest that BHHs belonging to Debbarm tribe of the Dakshin Dhumachhara, Dakshin Taidu and Pathaliaghat RPCs/villages have a higher score in comparison to the BHHs belonging to Chakma tribe living in Depachhara RPC/village.

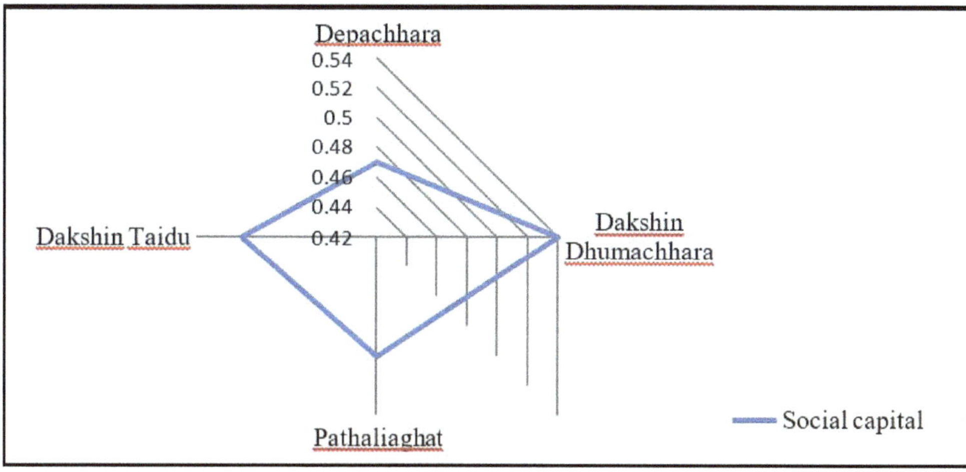

Fig. 5.2: Composite index score of social asset/capital of the BHHs of the RPCs.

Natural asset/capital

Natural asset/capital is extremely important as our livelihoods are dependent on its access. The indicators for measuring natural asset/capital are- size of landholding, access to land for *jhum* cultivation, access to common water resources (Pond/Steam/River) and access to common forest resources. In regard to natural asset/capital, the BHHs of Dakshin Dhumachhara RPC/village (figure 5.3) is found to be slightly better off than other RPCs/villages. With a composite index score of 0.46, the natural asset/capital of Dakshin Dhumachhara RPC/village is closely followed by Dakshin Taidu which has a composite index score of 0.44, Depachhara RPC/village (0.38 score) and Pathaliaghat RPC/village has the least composite index score of 0.37 (appendix 12). High access to landholding size between 1 - 1.5 ha (80.00 %) and access to water sources like Pond/Stream/River (60.00 %) by the BHHs of Dakshin Dhumachhara RPC/village contributes to the high composite index score of natural asset/capital of the BHHs living in Dhumachhara RPC/ village in comparison to the other three RPCs/villages. Field data suggest that the BHHs belonging to Debbarma tribe living in Dakshin Dhumachhara and Dakshin Tiaidu RPCs/villages reports of having better access to natural asset/capital than the Chakma tribe living in Depachhara RPC/village. It is interesting to note that although the BHHs of Pathaliaghat RPC/village belong to Debbarma tribe it scores the least in the composite index score of social asset/capital.

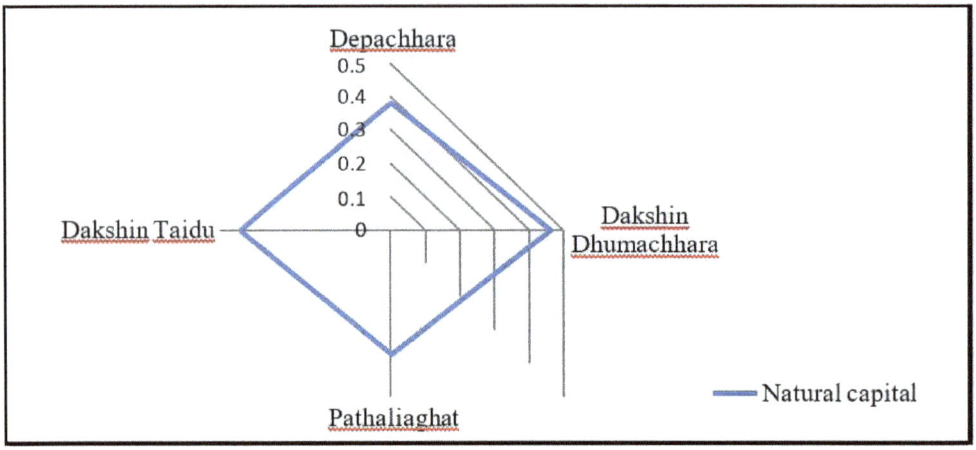

Fig. 5.3: Composite score of natural asset/capital of the BHHs of the RPCs.

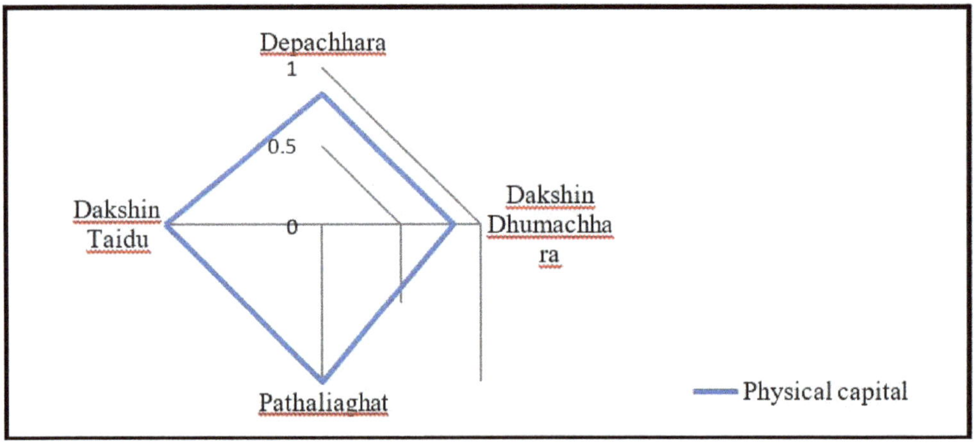

Fig. 5.4: Composite index score of physical asset/capital of the BHHs of the RPCs.

Physical asset/capital

Physical asset/capital is essential as without adequate access to proper infrastructures livelihoods are affected and this has a detrimental effect on the other four livelihood assets/capitals. The indicators to measure physical asset/capital are access to rubber processing centre within the RPC/village, access to blacktop road within a kilometre (km), access to market within 5 km, access to safe drinking water sources and sanitation within a meter from their respective residence, access to school within the RPC/village and access to electricity within their respective residence. Physical asset/capital of the four RPCs/villages (figure 5.4) suggests that the BHHs from Pathaliaghat and Dakshin Taidu RPCs/villages are better off in physical asset/capital than Dakshin Dhumachhara and Depachhara RPCs/villages. The composite index score of the four RPCs/villages indicates that Pathaliaghat and Dakshin Taidu RPCs/villages has the highest score (1) this is closely followed by Dakshin Dhumachhara and Depachhara RPCs/villages with the composite index score of 0.83 respectively. 100 percent of the BHHs living in Pathaliaghat and Dakshin Taidu RPCs/villages have access to rubber processing centre, access to blacktop road within a kilometre (km) distance, access to market within 5 km, access to safe drinking water source and sanitation, access school and access to electricity (appendix 13) in comparison to the BHHs of Dakshin Dhumachhara and Depachhara RPCs/villages. Figure 5.4 suggest that the BHHs belonging to the Debbarma tribe from Pathaliaghat abd Dakshin Taidu RPCs/villages are better off than the Chakma tribe from Depachhara RPC/village.

Financial asset/capital

Financial asset/capital is the most versatile of the five categories of assets/capitals as it paves in the way to enhance the other four assets/capitals. In fact, it is the asset/capital which often the poor lacks and thereby suffers from poverty and low-socio-economic development. The economic condition and the standard of living of a household are largely dependent on the income earned. Consequently the indicators for measuring financial asset/capital are the percentage share of households reporting different categories of monthly income. In regard to financial asset/capital as depicted in figure 5.5, the BHHs of Dakshin Taidu RPC/village are having the highest composite index score (0.45). followed by Dakshin Dhumachhara RPC/village (0.38 composite index score), Pathaliaghat RPC/village with a composite index score of 0.31 while Depachhara RPC/village reports the least in financial asset/capital with a composite index score of 0.31 (appendix 14). Higher share of BHHs in high income group come from Dakshin Taidu RPC/village as 33 percent of the BHHs of Dakshin Taidu RPC/village earn an income ranging between Rs 9,001 – 12,000 per month, 18.18 percent earned income between Rs 12,001 – 15,000 per month and 1.82 percent of the BHHs of Dakshin Taidu RPC/village earn income between Rs 15,001 – 18,000 per month (appendix 14). Field data suggest that the BHHs belonging to Debbarma tribe of Dakshin Tiaidu, Dakshin Dhumachhara and Pathaliaghat RPCs/villages are having higher financial asset/capital than the Chakma tribe of Depachhara RPC/village.

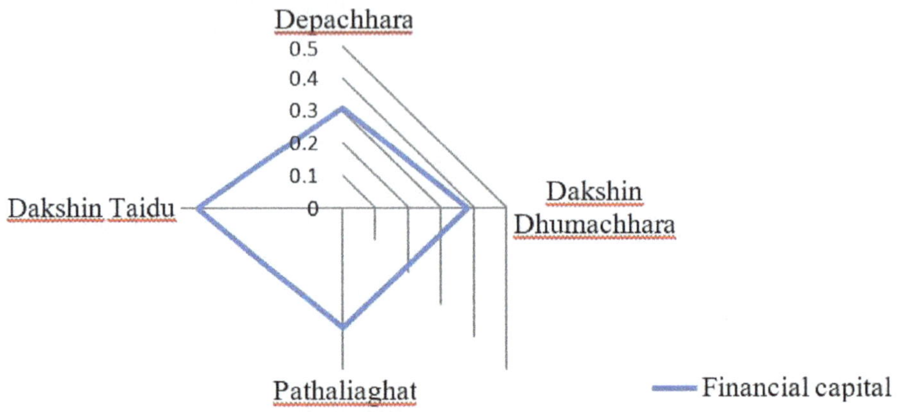

Fig. 5.5: Composite index score of financial asset/capital of the BHHs of the RPCs.

The composite index score of the four RPCs/villages has been further diagrammatically represented by livelihood asset pentagon as depicted in figure 5.6 in order to visualize and compare the strength and weakness of the five assets/ capitals of the four RPCs/villages.

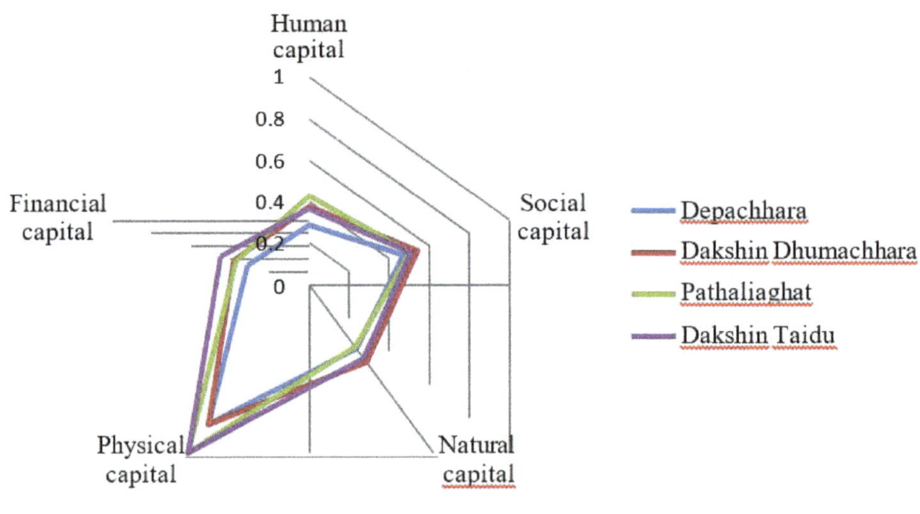

Fig. 5.6: Asset pentagon of five assets/capitals of the BHHs of the four RPCs.

Figure 5.6 suggests that physical assets/capital of all the four RPCs/villages have the best composite index score for physical asset/capital. The figure also suggests that social, financial and natural assets/capitals in all the RPC/village are similar with slight variation. Human asset/capital has also improved but in comparison to other assets/capitals it is lacking behind. It is observed that due to rubber plantation-based rehabilitation programmed the socio-economic conditions of the four RPCs/villages have enhanced as a result of access to rubber processing centre, increase in landholding size, access to training and capacity development activities and participation in programme related to rubber, increase in income from rubber plantation and modern skills associated with rubber. The figure also suggest that all the five assets/capitals of the BHHs belonging to the Debbarma tribe living in Dakshin Taidu, Pathaliaghat and Dakshin Dhumachhara RPCs/village have a higher score in comparison to the BHHs belonging to the Chakma tribe of the Depachhara RPC/village.

5.3 Socio-economic characteristics of the BHHs and NBHHs

5.3.1 Tribe composition

As already mentioned in Tripura there are nineteen ST communities and the major ST communities are Tripura/Tripuri/Tippera, Reang, Jamatia, Chakma, Halam and Mog. In the selected four RPCs/villages, field data suggests that the TRPCL has rehabilitated mainly three tribe's viz. Debbarma, Reang and Chakma. Among the BHHs the Debbarma tribe has the highest share (about 69 percent of the total BHHs) followed by Chakma (about 26 percent) while Reang tribe has a marginal share of only 5.24 percent of the total BHHs (table 5.2). The NBHHs also follow a similar pattern where the Debbarma is the most dominant group followed by the Chakmas and the Reangs and this is attributed to the fact that the NBHHs have been selected from the same villages.

Table 5.2: Distribution of tribe of the BHHs and NBHHs

Tribe	Share (in %) of tribe of the BHHs (n=210)	Share (in %) of tribe of the NBHHs (n=101)
Debbarma	69.05	66.34
Reang	5.24	9.90
Chakma	25.71	23.76

Source: Field survey, 2016.

1.3.2 Age and sex composition

The understanding of population structure becomes essential to examine and measure the amount of labour force available. The 210 surveyed BHHs consists of 986 people, of this 504 are male (51 %) and 482 are female accounting for 49 percent of the total population. About 68 percent of the total population belong to the working age group (15 – 59 years) while 32 percent constitute dependent population (below 15 years and 60 years and above age groups). Of the total working age group population, the share of male and female is almost same accounting for about 50 percent share each respectively. It is interesting to note that dependent population in the age group of below 15 years the male exceeds female population while the dependent population in the older age group of 60 years and above, female populations exceeds male population. On the other hand, out of the 101 NBHHs surveyed, the total population is 491 persons out of which about 59 percent of the total belongs to the working age group (15 – 59 years) while 41.43 percent are dependent population (below 15 and 60 years and above age groups). Of the 59 percent working age group population of the NBHHs about 52 percent are male population and about 48 percent are female (table 5.3). These suggest that the BHHs have slightly higher share of working age group population in comparison to the NBHHs.

Table 5.3: Share of population by age and sex (BHHs and NBHHs)

	BHHs (n=210)			NBHHs (n=101)		
Age group	**Total population**	**Male population**	**Female population**	**Total population**	**Male population**	**Female population**
0 - 14	271 (27.49)	149 (54.98)	122 (45.02)	179 (36.45)	96 (53.63)	83 (46.37)
15 – 59	670 (67.95)	333 (49.70)	337 (50.30)	288 (58.66)	151 (52.43)	137 (47.57)
60 +	45 (4.56)	22 (48.89)	23 (51.11)	24 (4.89)	10 (41.67)	14 (58.33)
Total	986 (100)	504 (51.12)	482 (48.88)	491 (100)	257 (52.34)	234 (47.66)

Source: Field survey, 2016 (figure in parenthesis indicate percentage of population to its respective total population).

It is found that of the 670 working age group (15 – 59 years) population of the BHHs, about 292 persons (43.58 percent) are found actively engaged in economic activities and earning income in cash, of this about 90 percent are male and only 10 percent are female population. On the other hand, of the 288 working age group (15 – 59 years) population of the NBHHs, about 116 persons (40.23 percent) are found engaged in economic activities, of this about 76 percent are male and about 24 percent are female population (table 5.3). This suggest that male population are more active as the bread earner than the female population in both the BHH and NBHHs, however female population from the NBHHs have a higher share in economic activities resulting to earning of income in cash in comparison to the BHHs.

5.3.3 Head of the households

Head of the household is usually defined as the member who initiates and undertakes in guiding major decisions of the household. About 99 percent of the BHHs are found to be male-headed while only 1 percent of the BHHs are found to be female-headed household (table 5.4). This suggests that male-headed households are more common in rural areas among the ST population of Tripura. Among the NBHHs also male-headed households are found more than the female-headed households although female-headed households are found slightly higher among the NBHHs than the BHHs.

Table 5.4: Head of the household by sex of the BHHs and NBHHs

Head of the household	Share (in %) of head of household of the BHHs (n=210)	Share (in %) of head of household of the NBHHs (n=101)
Male-headed households	98.57	97.03
Female-headed households	1.43	2.97

Source: Field survey, 2016.

1.3.4 Family type

As depicted in table 5.5, it is interesting to note that about 86 percent of the BHHs are nuclear families while the share of joint families is only 14 percent suggesting that even in rural areas the joint family system is slowly changing to nuclear family type. The same characteristic is also found for the NBHHs.

Table 5.5: Family profile of the BHHs and NBHHs

Family type	Share (in%) of family type of the BHHs (n=210)	Share (in %) of family type of the NBHHs (n=101)
Nuclear	86.19	87.13
Joint families	13.81	12.87

Source: Field survey, 2016.

5.3.5 Size of the households

As depicted in table 5.6, about 55 percent of the BHHs are having an average family size of about 6 members while 44 percent of the BHHs are having around 4 members. The total average size of the BHHs is about 5 members suggesting that the BHHs are following the norms of small family brought about by the family welfare programme. Field data suggests that the BHHs needs to hire labour to maintain their rubber plantation thereby providing livelihood opportunities to the NBHHs living in the area. The NBHHs also have a similar average family size of about 5 members. This suggest that family planning measures are working in the surveyed four RPCs/villages as there is shrinkage in size of the family.

Table 5.6: Share of BHHs and NBHHs as per size of family

Size group	BHHs (n=210)				NBHHs (n=101)			
	No. of house hold	Share in %	Total popul ation	Average size	No. of househ old	Share in %	Total popul ation	Average size
< 5	93	44.29	324	3.5	39	38.61	144	3.7
5 – 8	115	54.76	643	5.6	59	58.42	319	5.4
9 – 12	2	0.95	19	9.5	3	2.97	28	9.3
Total	210	100	986	4.7	101	100	491	4.9

Source: Field survey, 2016.

1.3.6 Literacy and levels of education

Literacy is defined as "the ability of a person to read and write at least simple message with some amount of understanding in any language". From 1991 India census onwards, the population in the early age groups of up to six years are ignored while calculating literacy rate (Hassan, 2015). In this research as well, the population of up to age group of six years are ignored while calculating literacy rates of both the BHHs and NBHHs.

About 77 percent of the population from the BHHs are found to be literate of which 55 percent are male and 45 percent are female population (table 5.7) suggesting a male-female gap of 5 percent in literacy rate. In all the age groups, male literacy exceeds the female literacy except in the age group of 15 – 24 years where female literacy (54 %) are found more than male literacy (46 %). High literacy rate is found in the age group of 7 – 44 years while low level of literacy is observed in the older age groups i.e. above 45 years. These suggest that the younger generation of the BHHs are more literate than the older generation as the access to education and value of education is recent here.

Table 5.7: Distribution of literacy rate as per age group of the BHHs and NBHHs

Size group	BHHs			NBHHs		
	Total literate	Male literate	Female literate	Total literate	Male literate	Female literate
7 – 14	196 (27.84)	105 (57.57)	91 (46.43)	118 (40.83)	66 (55.93)	52 (44.07)
15 – 24	142 (20.17)	65 (45.77)	77 (54.23)	72 (24.91)	39 (54.17)	33 (45.83)
25 – 34	167 (23.72)	89 (53.29)	78 (46.71)	46 (15.92)	22 (47.83)	24 (52.17)
35 – 44	123 (17.47)	64 (52.03)	59 (47.97)	29 (10.04)	16 (55.17)	13 (44.83)
45 – 54	54 (7.67)	41 (75.93)	13 (24.07)	12 (4.15)	8 (66.67)	4 (33.33)
55 – 64	17 (2.42)	15 (88.24)	2 (11.76)	8 (2.77)	5 (62.50)	3 (37.50)
65 +	5 (0.71)	5 (100)	0	4 (1.38)	3 (75.00)	1 (25.00)
Literate	704 (77.28)	384 (54.55)	320 (45.45)	289 (67.21)	159 (55.02)	130 (44.98)
Illiterate	207 (22.72)	76 (36.71)	131 (63.29)	141 (32.79)	68 (48.23)	73 (51.77)
Total	911	460	451	430	227	203

Source: Field survey, 2016 (figure in parenthesis indicate percentage share of literate and illiterates to its respective total population).

On the other hand, about 67 percent of the population of the NBHHs are found to be literate. Among the NBHHs also it is observed that the share of literate male population is more in comparison to the female literacy rate (table 5.7) and also literacy rate is higher for the younger generation when compared to the older generation. Consequently, the literacy composition of the BHHs and NBHHs is similar with marginal differences.

Field data suggests that of the total 77 percent literate population of the BHHs, only 267 peoples (37.93 %) during field survey are pursuing education at different levels. Out of this 54 percent are male and 46 percent are female students suggesting a male-female gap of 10 percent (table 5.8). This table further suggests that a large number of students (43 %) are found in the primary stage of education where the share of male children is as high as 59 percent in comparison female students (41%). The table also suggest the levels of education of the members of both the BHHs and NBHHs have marginal differences. Further the share of higher level of education reduces as one move from the primary level to the undergraduate level. This suggests that literacy and attainment of the different levels of education is recent here.

Table 5.8: Levels of education of the BHHs and NBHHs

Levels of education	BHHs			NBHHs		
	Total student	Male student	Female student	Total student	Male student	Female Student
Primary stage (classes I – V)	115 (43.07)	68 (59.13)	47 (40.87)	64 (45.07)	36 (56.25)	28 (43.75)
Middle stage (classes VI – VIII)	73 (27.34)	38 (52.05)	35 (47.95)	49 (34.51)	23 (46.94)	26 (53.06)
Secondary stage (classes IX – X)	71 (26.59)	35 (49.30)	36 (50.70)	27 (19.01)	16 (59.26)	11 (40.74)
Higher Secondary stage (classes XI – XII)	4 (1.50)	2 (50.00)	2 (50.00)	2 (1.41)	2 (100)	0
Undergraduate stage	4 (1.50)	2 (50.00)	2 (50.00)	0	0	0
Total	267 (100)	145 (54.31)	122 (45.69)	142 (100)	77 (54.23)	65 (45.77)

Source: Field survey, 2016 (figure in parenthesis indicate percentage of student to its respective total population).

1.3.7 Access to amenities, house type and ownership of consumable commodities

In order to assess the economic condition of the surveyed households of both the BHHs and NBHHs some proxy indicators have been selected to ascertain the economic condition, consequently information on access to amenities like electricity, source of drinking water, type of toilet facilities, house type (building material and ownership status of the house they live in) as well as possession of listed durable goods has been collected.

Field data (table 5.9) suggest that both the BHHs and NBHHs have 100 percent access to electricity suggesting that electricity has reached the rural areas of the state. Source of drinking water for the rural households is important as unprotected source of water can cause potentially fatal diseases like cholera, dysentery, typhoid etc. Therefore, protected source of water viz. protected wells, deep boreholes and hand pump are necessary to avoid such water borne diseases. It is found that most of the BHHs have access to safe drinking water sources with 41 percent using protected well, 32 percent using both hand pump and 27 percent using hand pump. It is observed that about 59 percent of the BHHs are using both pot and filter for storing water for drinking and kitchen use, while 28 percent uses filter for drinking water and 13 percent uses pots for storing drinking water. On the other hand, large number of the NBHHs (55 %) uses protected well as their source of drinking water which is higher share in comparison share to the share of BHHs. The share of well and hand pumps as the main source of drinking water among the NBHHs are slightly lower than the BHHs. It is observed that about 54 percent of the NBHHs use both pot and filter for storing water for drinking and kitchen use, while 26 percent uses filter for drinking water only and 20 percent uses pot for storing drinking water.

The use of traditional pit latrines is common among both the BHHs and NBHHs accounting for 100 percent respectively suggesting that modern sanitation facilities are not yet popular among both the sets of surveyed households.

Table 5.9: Household amenities of the BHHs and NBHHs

Amenity	Share (in %) of access to amenities of the BHHs (n=210)	Share (in %) of access to amenity of the NBHHs (n=101)
Electricity	100	100
Source of drinking water within 500 meters		
Hand pump	27.14	24.75
Well	40.95	55.44
Both hand pump and well	31.91	18.81
Total	100	100
Sanitation facility		
Traditional pit toilet	100	100
Building material of the house		
Mud wall with tin roof	93.33	87.13
Half mud wall and bamboo/tin wall with tin roof	5.24	12.87
Brick wall with tin roof	1.43	0
Total	100	100
Cooking fuel		
Firewood, straw	100	100

Source: Field survey, 2016.

The type of building material used generally indicates the economic condition of the households. In rural areas of Tripura, mud wall with tin/thatch roof is common, tin wall with tin/thatch roof, half mud with tin/bamboo wall and tin/ thatch roof etc are considered secure houses for living. In this context, about 93 percent of the BHHs are found living in secure houses consisting of mud wall with tin roof, while only 5 percent of the BHHs have half mud wall and bamboo/ tin wall with tin roof and 2 percent of the BHHs have houses with brick wall with tin roof. Among the NBHHs about 87 percent of the households are found living in mud wall with tin roof house and 13 percent of the NBHHs live in mud wall and bamboo/tin wall with tin roof, while brick wall with tin roof house is not found among the NBHHs. These suggest that the both the BHHs and the NBHHs live mostly in houses having mud walls with tin roofs suggesting similar living conditions except for a small 2 percent of the BHHs who prefer to live in houses having brick wall with tinned roofs usually suggesting better economic condition (table 5.9).

Various type of cooking fuel used by a household reflects both economic status as well as exposure to varying types of pollutant. Both the BHHs and NBHHs are found to be using firewood or straw as their main source of fuel for cooking purpose (table 5.9).

Information pertaining to the ownership of a list of consumable items from both the BHHs and NBHHS has been collected to assess their economic condition and their immediate coping strategies to meet financial crises when these commodities can be sold out to meet small emergencies. Ownership of a radio or television is a measure of access to mass media; ownership of a bicycle, motorcycle or car reflects means of transport, which can be important during emergencies and daily uses. Ownership of a telephone or cell phone indicates easy communication with other users from a distance. All the BHHs and NBHHs own a paraffin lamp as in the surveyed localities interruption of electricity supply is frequent. About 85 percent of the BHHs own a cell phone while 77 percent own a television and only 29 percent own a radio. About 69 percent of the BHHs own table and chairs while only 10 percent own a sofa set. Bicycles are the most common type of vehicle owned by the BHHs, where 77 percent of the households have a bicycle. Ownership of motorcycle and car are found rare among the BHHs; 2 percent of the respondents from BHHs reports owning both motorcycle and bicycle while only 1 percent of the respondent report owning a car. Similar patterns of possession of consumable items are also observed for the surveyed NBHHs suggesting a similar economic condition with that of the BHHs (table 5.10).

Table 5.10: Possession of consumer items of the BHHs and NBHHs

Durable goods	Share (in %) of durable goods of the BHHs	Share (in %) of durable goods of the NBHHs
Households goods		
Paraffin lamp	100	100
Radio	29.05	22.77
Television	76.67	53.46
Cell phone	85.24	74.26
Sofa set	10.00	3.96
Table and chairs	69.05	58.42
Means of transport		
Bicycle	76.67	67.33
Motorcycle	2.38	0.99
Car	0.95	0
Both bicycle and motorcycle	2.28	0

Source: Field survey, 2016.

1.3.8 Household income

Field data suggests that the main source of income of the BHHs is from rubber plantation. The approximate monthly income of the surveyed BHHs from rubber plantation ranges between Rs 3,600 to Rs 14,000. While that of the NBHHs ranges from Rs 3,000 to Rs 8,000. The monthly income of both the BHHs and NBHHs has been categorized into four classes viz. very high income group (monthly income group ranging from Rs 12,001-15,000), high income group (monthly income ranging from Rs 9,001-12,000, middle income group (monthly income group ranging from Rs 6,001-9,000) and the low income group (monthly income ranging between Rs 3,000- 6,000). Table 5.11 shows the distribution of income of the BHHs and NBHHs.

Table 5.11 indicates that most of the NBHHs fall within low income group (40.59 %) followed by medium income group (7.92 %) in comparison to the BHHs who have better income status as only 6 percent of the respondents report low monthly income followed by 55.24 percent of the respondents reporting medium monthly income, while 35.24 percent of respondents here report high monthly income which is totally absent for the NBHHs set. Another 3.33 percent of the respondents from the BHHs report a relatively very high income of Rs 9,001-12,000 per month from rubber plantation. The respondents from BHHs report that they earn their income mainly from selling latex and also from tapping the rubber trees. This has created steady flow of income from rubber plantation.

On the other hand the NBHHs reports that they earn income as tappers and latex collectors and as rubber processors in the rubber processing centres. This suggests that rubber plantation has generated employment opportunities; it is observed that 48.51 percent of the NBHHs (table 5.11) are directly earning income from activities associated with rubber plantation.

Table 5.11: Distribution of monthly income groups derived from rubber plantation (BHHs and NBHHs)

Income group	Share of income (in %) of the BHHs (n=210)	Share of income (in %) of the NBHHs (n=101)
Low (Rs 3,000 – 6,000)	6.19	40.59
Medium (Rs 6,001 – 9,000)	55.24	7.92
High (Rs 9,001 – 12,000)	35.24	0
Very high (Rs 12,0001 – 15,000)	3.33	0

Source: Field survey, 2016.

Apart from rubber plantation, the BHHs are also found earning income from other non-farm activities like shop keeping, drivers, carpentry, petty business in weekly market, working as mechanic and party decorators (mainly seasonal in nature). Similarly, respondents from NBHHs also reports the same with the further addition of casual labour force comprising 51.49 percent of the total (table 5.16) which is absent among the BHHs (table 5.15). At present, members from both BHHs and NBHHs enjoy income from MGNREGA as they are assured

100 days employment in a year. These suggest that rubber plantation to some extent has changed the livelihood practices of the people here as a large chunk of the respondents from the BHHs reports livelihood practices derived from the plantation itself along with other non-farm activities, which were totally absent before the plantations were introduced here. It is interesting to note that environmental income derived from forest has reduced with coming up of the rubber plantation.

Field data (table 5.12) suggests that most of the NBHHs are involved in earning cash income from diversified sources in comparison to the BHHs. About 12 percent and 4 percent of the BHHs respectively earn income of below Rs 6,000 per month and Rs 6,001-9,000 per month respectively from non-farm activities not directly associated with rubber plantation. On the other hand, about 65 percent of the NBHHs reports that they earn monthly income of below Rs 6,000 and have diversified sources of livelihoods not directly related to rubber plantation.

Table 5.12: Distribution of monthly income groups derived from other sources (BHHs and NBHHs)

Income group	Share of income (in %) of the BHHs (n=210)	Share of income (in %) of the NBHHs (n=101)
Below 6,000	12.38	65.35
6,001 – 9,000	3.81	4.95

Source: Field survey, 2016.

The above table 5.11 and 5.12 clearly suggests that although the NBHHs are involved in activities associated with rubber plantation for earning income but most of the NBHHs are involved in other activities for earning income in comparison to the BHHs. The main reasons for this low involvement in other activities among the BHHs are due to regular flow of income from rubber plantation to these households. This suggests that rubber plantation have resulted to a shift from traditional livelihood practices to modern ones based on rubber plantations further the income generated from the environment and farm activities for both the BHHs and NBHHs living in the same ecological set up are becoming less attractive.

Field data suggests that both the BHHs and NBHHs are rearing livestock such as poultry/hen, pigs, goats, duck and cows which are mainly reared for self-consumption with little surplus to be sold in the market. The livestock acts as stock to meet the cash demand during periods of financial crises, thereby helping in coping strategy. About 78 percent of the BHHs reports rearing of livestock in comparison to the NBHHs (69 %) thus the BHHs have a slightly higher share of livestock rearing when compared to the NBHHs. This can be attributed to the fact that the BHHs have better flow of income in comparison to their counterpart (NBHHs).

1.4 Major livelihood practices of the BHHs and NBHHs

In this section, the livelihood practices, livelihood diversification and constraints to livelihood diversification of the two sets of surveyed households i.e.

the BHHs and NRHHs are discussed, analysed and compared in order to assess the role of rubber plantation in diversifying livelihood practices of the BHHs and NBHHs.

1.4.1 Livelihood practices

Livelihood refers to the diverse sources of income either in cash or kind which the poor derive from both farm and non-farm activities including the physical environment to eke out a living. In this study major livelihood practices have been identified in terms of income generating activities. A report from World Resource Institute (2005) suggest that in most of the developing economies the marginalized households and also individuals are involved in different economic activities to eke out a living and also improve on their income generation capacity. Households with diverse livelihood practices differ from households who are involved on a single full- time specialized livelihood activity. The number of income sources by both the BHHs and NBHHs has been assessed measured and compared in order to examine the role of rubber plantation in livelihood diversification and changing livelihood option.

It is found that tapping and latex collection associated with rubber plantation are the major livelihood practice of the BHHs. However, about 82 percent of the respondents from the BHHs reports only tapping and latex collection as their major livelihood practice while about 18 percent of the BHHs are found practising other livelihood activities besides the livelihood activities associated with rubber plantation (table 5.13). Besides tapping and latex collection associated with rubber plantation, other livelihood activities of the BHHs are shop keeping, working as drivers, petty business in the weekly market, working as mechanic, carpentry and party decorators. About 9 percent of the respondents from the surveyed BHHs reports rubber plantation along with shop keeping as their major livelihood practice, 5 percent of the respondents from this set, reports both rubber plantation and working as drivers as their main livelihood practices, while 2 percent each of the respondents report both rubber plantation and carpentry followed by rubber plantation and petty business in the weekly market, working as mechanic and party decorators as their major livelihood practices (table 5.13). This suggests that the BHHs have been able to set up various earning sources. The respondents from the BHHs report that various income generating activities have been set up only after being rehabilitated through the rubber plantation programme. They are of the opinion that the flow of income from rubber plantation have allowed them to save money which they have later invested to set up petty business such as opening of shops in their respective villages and nearby weekly market, learning new skills like driving, learning mechanical skills, carpentry etc and setting up of small business, as they can afford to buy the necessary carpentry tools etc.

For the NBHHs it is found that tapping of latex and its collection, processing of latex and smoking of rubber sheets, as well as employment as guards in the rubber processing centre are the new livelihood practices associated with introduction of rubber plantation. About 20 percent of the respondents from the NBHHs reports tapping and collection of latex as their major livelihood practice, while 9 percent of the respondents reports processing of latex and smoking of rubber sheets as their major livelihood practice (table 5.14). It is interesting to note that out of the 20 percent, who reported tapping and collection of latex as their major livelihood practice, 82.35 percent are found to be engaged as tappers in private rubber plantation and only 17.65 percent are engaged as tappers in rubber plantation managed by the TRPCL. As depicted in table 5.14, only 1 percent of the respondent from NBHHs reports that employment as guard in the rubber processing centre as the major livelihood practice. It is also noticed that members of NBHHs also report different non-farm livelihood practices like shop keeping, driving, business in weekly market, working as mechanic, carpentry and party decorators besides tapping and collection of latex, processing of latex and smoking of rubber sheets associated with rubber plantation. However, most of the NBHHs (about 52 percent) report that their major livelihood practice is working as casual labour.

As noticed from table 5.14, all the respondents from the NBHHs who are practicing non-farm activates like shop keeping, working as drivers, small business in weekly market, working as mechanic, carpenter and party decorators are from the same households who report livelihood practices associated with rubber plantation like tapping and collection of latex (TCL), processor of latex and smoking of rubber sheet (PLSRS). The respondents from these households report that the income generated from rubber plantation have increased their purchasing power and savings consequently over time they are in a position to invest in small business like opening up of grocery shops, learning new skills like driving, learning mechanical jobs, setting up of small business and purchasing necessary carpentry tools, decoration materials etc which has helped them not only to set up petty business thereby increasing their income level but also has helped them in generating small jobs where the other villagers can get small jobs.

Field data suggests that 38 numbers (18.10 %) of the total respondents from the BHHs and 19 numbers (18.81 %) of the total respondents from the NBHHs report their major livelihood practice as non-farm activities along with livelihood practices associated with rubber plantation. This suggests that members from both the BHHs and NBHHs are generating income from non-farm activities thereby there is a shift in livelihood practices of the population living here with the introduction of rubber plantation.

It is interesting to note from Focus group discussion (FGD) during field survey as well as from survey of households (BHHs and NBHHs) that there has been a prior understanding between the TRPCL officials and the villagers, that those families who have not been covered under the rehabilitation programme through rubber plantation within a particular village/RPC will find employment as processor of latex and smoking of rubber sheets. It is with this understanding that about 9 percent of the respondents from the NBHHs reports that their major livelihood practices are latex processing and smoking of rubber sheets in the rubber processing centre developed by the TRPCL in the surveyed villages/RPCs. It is also interesting to note that a person who is employed as a guard in the rubber processing centre has been employed in lieu of exchange of land for construction of the rubber processing centre in his land. This suggests that rubber plantation has not only shifted the livelihood practices of the erstwhile *jhumias*, marginalized farmers and landless peoples (now BHHs) to tappers but has also generated employment opportunities for the members of the households (NBHHs) who are not directly rehabilitated under the state government's rubber plantation based rehabilitation programme.

1.4.2 Livelihood diversification

In order to asses, analyze and compare the livelihood diversification related to livelihood practices of both the BHHs and NBHHs, the modified version of Simpson index of diversity (SID) has been used. As mentioned in the methodology section, SID value ranges between 0-1 and value closer to score of 0 represents less diversity and indicates complete specialization while the SID value closer to 1 represents higher diversification in livelihood practices.

Table 5.13: Income sources and livelihood diversification of the BHHs

Major Livelihood practices	No. of BHHs
Rubber plantation (tapping and collection of latex)	172 (81.90)
Rubber plantation and shop keeping	19 (9.05)
Rubber plantation and working as driver	11 (5.24)
Rubber plantation and carpenter	4 (1.90)
Rubber plantation and petty business in local market	2 (0.95)
Rubber plantation and working as mechanic	1 (0.48)
Rubber plantation and party decorator	1 (0.48)
Simpsons Index Diversity value	0.32

Source: Field survey, 2016 (figure in parenthesis indicate percentage share).

The SID value of the BHHs is 0.32. This suggests relatively low livelihood diversification of income sources as the BHHs are found engaged in seven incomes generating activates. Most of the BHHs (about 82 %) are found earning their income from rubber plantation alone as it gives a steady source of income. Only 18.10 percent of the BHHs are found engaged in others income generating activates mainly from rubber plantation and owning of shops, working as drivers, carpenter, mechanic and party decorator along with business in weekly markets (table 5.13).

The SID value of the NBHHs is 0.68. This suggests higher diversity in livelihood practices as the NBHHs are found engaged in twelve income generating activities (table 5.14). Among these activities, about 51.49 percent of the respondents from the NBHHs reports that their major livelihood practice as casual labour followed by tapping and collection of latex (T&CL), processing of latex and smoking of rubber sheet (PLSRS). Other sources from where the NBHHs earn their major income are shop keeping, engaged as drivers, carpentry, petty business in weekly market, mechanic and as party decorators.

The SID value of the BHHs is 0.32 (table 5.13) in comparison to the SID value of NBHHs which is 0.68 (table 5.14) suggesting relatively lower diversification of livelihood practices among the BHHs in comparison to the NBHHs. This may be attributed to the fact that most of the BHHs have more steady income from the rubber plantations and are relatively financially well off when compared to the NBHHs living in the same village. On the other hand, as the NBHHs are having lesser income they need to diversify their source of income consequently diversification of livelihood practices is higher among the NBHHs in comparison to the BHHs.

Table 5.14: Income sources and livelihood diversification of the NBHHs

Major livelihood practices	No. of NBHHs
Casual labour (CL)	52 (51.49)
Tapper and latex collection (T&CL)	20 (19.80)
Latex processor and smoking of rubber sheet	9 (8.91)
Employed as guard in the RPC	1 (0.99)
T&CL and shop	4 (3.96)
T&CL and working as driver	3 (2.97)
TLC and carpentry business	2 (1.98)
TLC and petty business in local market	4 (3.96)
TLC and mechanic	1 (0.99)
LPSRS and driver	2 (1.98)
LPSRS and mechanic	2 (1.98)
LPSRS and party decorator	1 (0.99)
Simpson's Index of Diversity value	0.68

Source: Field survey, 2016 (figure in parenthesis indicate percentage share).

1.4.3 Constraints to livelihood diversification

Diversification of livelihood by the households is considered as an important survival strategy for the economically marginalized section of the society. However, there are certain constraints to livelihood diversification in the selected RPCs/villages which has been observed during field survey.

Some of the major constraints observed and reported by the respondents from both the sets of households are:

i) *Distance and lack of transport facilities-* Distance from the market or town with lack of regular transport flow are major constraints to livelihood diversification. This has impacted both buying and selling of goods and services as well as for employment in the non-farm activities away from the RPCs/villages.

ii) *Poor asset base-* Most of the surveyed households do not have much asset in the form of savings which acts as a deterrent to livelihood diversification. Poor asset base hinders the rural households to set up self-employment generating activities as well as improve their existing skills and learn new skills thus hindering increase in human capital.

iii) *Lack of credit facilities-* Lack of access to institutional credit acts as a deterrent in the process of livelihood diversification. In the absence of credit facilities from the institutional agencies, the households are not able to start new business or enterprises.

iv) *Lack of awareness and training facilities-* Apart from MGNREGA scheme, the households are unaware about others government schemes that are being provided for the development of rural areas. It must be mentioned here that no government initiative are taken up to generate awareness and interest among the villagers about various schemes available for rural development. There is total absence of any NGOs activities to generate awareness and inform the households about the different schemes which may help them to generate income.

v) *Lack of opportunities-* Besides rubber related activities there are not much non-farm jobs within and around the selected surveyed RPCs/villages.

vi) *Fear of taking risk-* Due to the poor asset base of the surveyed households of both the BHHs and NBHHs and lack of support from institutions and administration has made the people here less enterprising thus there is lack of entrepreneurship in all the four selected RPC/villages.

1.5 Comparative analysis of the five livelihood assets of the BHHs and NBHHs

As already mentioned earlier, in order to assess the impact of rubber plantation on the socio-economic conditions of the people and to evaluate whether livelihoods generated through rubber plantation programme are sustainable or not, the SLA has been used (DFID, 1999). Consequently, the five assets/capitals i.e. human, social, natural, physical and financial assets/capitals associated with livelihood studies has been discussed and measured with suitable indicators (mentioned in the methodology section of Chapter I) for both the BHHs and NBHHs. Further with the help of range equalization method (UNDP method of Human Development Index) (Ghosh et al., 2006) a scale free aggregate index score ranging between 0-1 representing each indicator of the five livelihood assets/capitals has been computed. Consequently, a composite index score of the five assets/capitals of both the BHHs and NBHHS has been measured, compared and diagrammatically represented by livelihood asset pentagon to assess the role of rubber plantation in enhancing socio-economic development and the level of sustainability of each livelihood asset/capital by comparing the score of each livelihood asset/capital to the theoretical norm.

1.5.1 Comparative analysis of the livelihood assets/capitals of BHHs and NBHHs

(i) Human asset/capital

Field data suggest that in case of percentage share of working population to total population, the BHHs have a higher share (67.95 %) in comparison to the NBHHs (58.66 %). The share of literacy rate is also higher among the BHHs (77.28 %) than the NBHHs (67.21 %) with a difference of 10.07 percent. However, in case of traditional skills known by both the members of the BHHs and NBHHs, it is interesting to note that the members of the NBHHs have retained these skills as 73.27 percent of the respondents from the NBHHs reports that some members within their household still practice and know such skills associated with making of handicraft like basketry, weaving, *jhum* cultivation etc against 68.10 percent of the respondents from the BHHs who reports the same. This suggests that with the introduction of rubber plantation the traditional skills among the BHHs are slowly waning out and is getting replaced by modern skills. Consequently 98.57 percent of the respondents from the BHHs reports that their family members know modern skills like tapping, mechanic work, decorating, driving etc whereas the share of members from NBHHs of knowing such modern skills are low as only 54.46 percent respondents from NBHHS report the same (table 5.15).

Table 5.15: Indicator suggesting human asset/capital of the BHHs and NBHHs

Indicator	Share (in %) of indicator from the BHHs (n=210)	0-1 scale value	Share (in %) of the indicator from the NBHHs (n=101)	0-1 scale value
Working population	67.95	0.00	58.66	0.22
Literacy rate	77.28	0.30	67.21	0.68
Traditional skills known by the members of the household	68.10	0.005	73.27	1.00
Modern skills known by the members of the household	98.57	1.00	54.46	0.00
Composite index score of human asset/capital		0.33		0.48

Source: Field survey, 2016.

(ii) Social asset/capital

It is observed from table 5.16 that 100 percent of the respondents from BHHs report of having good rapport with the TRPCL officials, while about 36.63 percent of the respondents from NBHHs report of sharing good relationship with the TRPCL officials in their respective RPC/village. The BHHs are found having higher share in case of training and capacity development activities (59.05 %) and participation in programme or meeting (89.05 %) than the NBHHs as only 23.76 and 27.72 percent report of attending training and capacity development activities and participation in programme or meeting respectively. In regards to the share of committee member in the locality, the NBHHs are having higher share (15.84 %) than the BHHs where the share as committee member in the locality is 12.38 percent. Both the BHHs and NBHHs are having almost the same share in case of membership in Self Help Group (SHG) although the NBBHs are slightly higher (table 5.16)

Table 5.16: Indicators suggesting social asset/capital of the BHHs and NBHHs

Indicator	Share (in %) of indicator of the BHHs (n=210)	0-1 score value	Share (in %) of indicator of the NBHHs	0-1 score value
Good relationship with TRPCL officials in their respective RPC/village	100	1.00	36.63	1.00
Training and capacity development activities	59.05	0.57	23.76	0.58
Committee member in the locality	12.38	0.07	15.84	0.32
Member in Self Help Group (SHG)	5.71	0.00	5.94	0.00
Participated in programme or meeting	89.05	0.88	27.72	0.71
Composite index score of social asset/ capital		0.50		0.52

Source: Field survey, 2016.

(iii) Natural asset/capital

Field investigation (table 5.17) suggests that the size of landholding varies between 0.4 ha to 2.5 ha which has been grouped into three classes i.e. Less than 1 ha, 1 to 1.5 ha, and above 1.5 ha. Field data suggest that the BHHs have higher landholding size than the NBHHs. It is observed from table 5.17 that about 68 percent of the BHHs have landholding size ranging between 1-1.5 ha and 13 percent of the BHHs have landholding size of above 1.5 ha while about 21 percent of the respondents from NBHHs reports having landholding size of 1-1.5 ha and only 7.92 percent of the respondents from NBHHs reports of having landholding size of above 1.5 ha. The large percentage share of BHHs having landholding size of 1-1.5 ha is mainly due to the rubber plantation based rehabilitation programme where each beneficiary has been allotted 1 ha of land for rubber plantation. On the other hand, there is a higher share of small size of landholding of less than 1 ha (71.29 %) among the NBHHs, only 18.57 of the respondents from BHHs report of having landholding size of less than 1 ha and it is to mentioned here that the BHHs are supposed to be allotted 1 ha of land for rubber plantation but as reported the said 1 ha of land for rubber plantation has not been allotted.

Table 5.17: Indicator suggesting natural asset/capital of the BHHs and NBHHs

Indicator	Share (in %) of indicator of the BHHs (n=210)	0-1 score value	Share (in %) of indicator of the NBHHs (n=101)	0-1 score value
Landholding size below 1 ha	18.57	0.22	71.29	1.00
Landholding size between 1 - 1.5 ha	68.10	1.00	20.79	0.20
Landholding size of above 1.5 ha	13.33	0.14	7.92	0.00
Access to land for *jhum* cultivation	4.29	0.00	10.89	0.05
Access to water sources (Pond/Steam/ River)	30.00	0.40	27.72	0.31
Access to common forest resource	55.24	0.80	60.40	0.83
Composite index score of natural asset/ capital		0.43		0.40

Source: Field survey, 2016.

It is interesting to note that only 4.29 percent of the respondents from BHHs report of having access to land for jhum cultivation suggesting that rubber plantation has reduced the practice of jhum cultivation. In case of the NBHHs, access to land for jhum cultivation is found slightly higher (10.89 %) in comparison to their counterpart (BHHs).

Access to common water resources (Pond/Steam/River) of both the BHHs and NBHHs are similar however the BHHs have slightly higher access to common water resources with a difference of just 2.28 percent.

About 55 percent of the respondents from BHHs reports of having access to common forest resource mainly for collecting firewood, food items and other materials for households needs while 60.40 percent of the respondents from NBHHs reports of having access to common forest resource.

(iv) Physical asset/capital

It is observed from table 5.18 that both the respondents from the BHHs and NBHHs reports of having 100 percent access to infrastructures facilities like blacktop road within a kilometre, safe drinking water sources (hand pump and well) and sanitation (traditional pit), access to school within the RPC/village and access to electricity. In case of access to rubber processing centre within the RPC/village the BHHs report 100 percent access while only 21.79 percent of the NBHHs report of having access to rubber processing centre. Here, the 21.79 percent of the NBHHs who report of having access to the rubber processing centre are those who are employed as processors of latex in the processing centre and as tapper and collector of latex in the rubber plantation developed by the TRPCL within the respective RPC/village. In case of access to market within 5 km, only about 26 percent of the BHHs report that the distance of the market from their locality is within 5 km while in case of NBHHs about 28 percent report that access to market is within 5 km suggesting that most of the households from both the BHHs and NBHHs have to travel more than 5 km to reach the nearest market.

Table 5.18: Indicators suggesting physical asset/capital of the BHHs and NBHHs

Indicator	Share (in %) of the indicator BHHs (n=210)	0-1 score value	Share (in %) of indicator of the NBHHs (n=101)	0-1 score value
Access to rubber processing centre within the respectiveRPC/village	100	1.00	20.79	0.00
Access to blacktop road within a kilometre (km)	100	1.00	100	1.00
Access to market within 5 km	26.19	0.00	27.72	0.09
Access to safe drinking water source and sanitation	100	1.00	100	1.00
Access to school within the RPC/village	100	1.00	100	1.00
Access to electricity in the house	100	1.00	100	1.00
Composite index score of physical asset/capital		0.83		0.68

Source: Field survey, 2016.

(v) Financial asset/capital

Field investigation suggests that most of the BHHs earn a higher income in comparison to the NBHHs. It is observed from table 5.19 that 52.38 percent of the respondents from the BHHs reports a monthly income ranging between Rs 6,001-9,000 and 30 percent of the BHHs are found earning income ranging between Rs 9,001-12,000 per month. On the other hand, about 25 percent of the respondents from the NBHHs reports of earning an income ranging between Rs 6,001-9,000 per month and only 3.96 percent of the NBHHs monthly income ranges between Rs 9,001- 12,000. Most of the respondents from NBHHs (69.31 %) earn an income ranging between Rs 3,000-6,000 per month while only 2.38 percent of the BHHs earn a monthly income ranging between Rs 3,000-6,000. None of the NBHHs are found earning incomes above Rs 15,000 per month while 1.43 percent of the respondents from BHHs reports earning a monthly income ranging between Rs 15,001-18,000. The higher income level per month of the BHHs when compared to the NBHHs are mainly due to high income flow from rubber plantation besides various other sources which are already discussed in table 5.11.

Table 5.19: Income group suggesting financial asset/capital of the BHHs and NBHHs

Income group	Share of income (in %) of the BHHs (n=210)	0-1 score value	Share of income (in %) of the NBHHs (n=101)	0-1 score value
Rs 3,000 – 6,000	2.38	0.02	69.31	1.00
Rs 6,001 – 9,000	52.38	1.00	24.75	0.36
Rs 9,001 – 12,000	30. 00	0.56	3.96	0.06
Rs 12,001 – 15,000	13.81	0.24	1.98	0.03
Rs 15,001 – 18,000	1.43	0.00	0	0.00
Composite index score of financial asset/capital		0.36		0.29

Source: Field survey, 2016.

1.5.2 Livelihood sustainability of the BHHs and NBHHs

To determine sustainability of the five livelihood assets/capitals of both the BHHs and NBHHs, composite index score of all the assets/capitals are computed and each asset/capital has been compared to the theoretical norm, in order to measure the level of sustainability. As per the theoretical norm on the level of sustainable livelihood developed by Kamaruddin and Samsudin (2014) the index score ranging from 0 to 1 are categorized into four groups i.e. 0 - 0.25, 0.26 - 0.50, 0.51 - 0.75 and 0.76 – 1. Based on the level of sustainable livelihood index range, the sustainability are classified into- 1 representing highly sustainable (score value ranging from 0.76- 1), 2 representing moderately high sustainable (score value ranging between 0.51- 0.75), 3 representing moderately sustainable (score value ranging between 0.26-0.50) and 4 representing low/un-sustainable (score value ranging from 0-0.25).

It is observed from table 5.20 that the composite index score of the BHHs range from as high as 0.83 score (physical asset/capital) to as low as 0.33 score (human asset/capital) while the composite index score of the NBHHs range from 0.68 score (physical asset/capital) being the highest to as low as 0.29 score (financial asset/ capital).

Table 5.20: Composite index score of the five assets/capitals of the BHHs and NBHHs

Asset/capital	Composite index score of the BHHs	Composite index score of the NBHHs	Gap in composite index score between the BHHs and NBHHs
Human asset/capital	0.33 (3)	0.48 (3)	-0.15
Social asset/capital	0.50 (3)	0.52 (2)	-0.02
Natural asset/ capital	0.43 (3)	0.40 (3)	0.03
Physical asset/ capital	0.83 (1)	0.68 (2)	0.15
Financial asset/ capital	0.36 (3)	0.29 (3)	0.07

Source: Computed by the researcher (figure in parenthesis represent level of sustainability).

Among the five assets/capitals, only physical asset/capital of the BHHs is found to be highly sustainable with a composite index score of 0.83 while the other four assets/capitals i.e. social asset/capital (0.50 score), natural asset/capital (0.43 score), financial asset/capital (0.36 score) and human asset/capital (0.33 score) are found to be moderately sustainable. On the other hand, among the five assets/capitals of the NBHHs, physical and social assets/capitals with 0.68 score and 0.52 score respectively are found to be moderately high sustainable while human, natural and financial assets/capitals with a score of 0.48, 0.40, and 0.29 respectively are found to be moderately sustainable (table 5.20).

The composite index score of three assets/capitals i.e. physical, financial and natural assets/capitals of the BHHs are higher in comparison to the NBHHs having a gap of 0.15, 0.07 and 0.03 respectively. The composite index score of two assets/capitals i.e. human and social assets/capitals of the BHHs are however found to be weak in comparison to the NBHHs. Composite index score of the human and social assets/capitals of the NBHHs are higher than the BHHs with a gap of 0.15 and 0.02 respectively (table 5.20).

The composite index score of the five assets/capitals of the BHHs and NBHHs is represented by livelihood asset pentagon (figure 5.7).

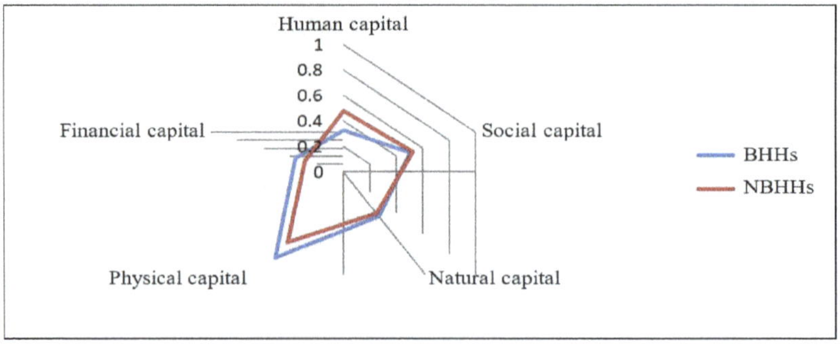

Fig. 5.7: Livelihood asset pentagon of the BHHs and NBHHs.

The asset pentagon clearly indicates that the BHHs are having better score in three assets/capitals i.e. physical, financial and natural assets/capitals in comparison to the score of the NBHHs. However, the NBHHs scores higher than the BHHs in case of human asset/capital and natural asset/capital when compared to the BHHs (figure 5.7).

The score value of physical asset/capital of the BHHs is higher when compared to the physical asset/capital score value of the NBHHs. This is due to the fact that the BHHs have better access to infrastructure facilities viz. access to rubber processing centre. The regular flow of income as well as higher share of high income group by the BHHs mainly due to rubber plantation are the main factors leading to higher composite index score in regards to financial asset/capital by the BHHs when compared to the NBHHs. The BHHs also have a higher composite index score for natural asset/capital when compared to the composite index score value of NBHHs, suggesting a gap of 0.03. This is due to the fact that the BHHS have a higher share in the bigger landholding size, as at least 1 ha of land is allotted to each BHH for rubber plantation based rehabilitation programme. On the other hand, the composite index score of social asset/capital of the BHHs are found slightly lesser in comparison to the composite index score of the NBHHs. This is due to the fact that the NBHHs have higher share in participation of training and capacity development activities and are also members of committees in the locality in comparison to BHHs. In regards to the composite index score of human asset/capital the BHHs scores less in comparison to the NBHHs. This is attributed to the fact that at present, substantial members from NBHHs have still retained their traditional skills when compared to the members from BHHs.

1.6 The rubber plantation-based rehabilitation programme and its outcome

As already mentioned rubber plantation here has been introduced to ensure sustainable livelihood and reduction of poverty among the marginalized section and field data suggests that the TRPCL has been successful in enhancing the five livelihood assets/capital of the population living in the four surveyed RPCs/villages. An analysis of the five livelihood assets/capitals suggest that the BHHs have gained in physical asset/capital, natural asset/capital especially access to land, financial asset/capital especially regular flow of income as well as higher share of the BHHs in higher income groups, social asset/capital and human asset/capital like learning new and modern skills like tapping, collection of latex etc.

It is observed that the livelihoods of the poor households are usually built on different assets/capitals thus access to these assets/capitals are considered as the major challenge for the poor households, consequently allocation of land to each BHHs for rubber plantation has not only provided flow of income to the BHHs but has helped in replenishing degraded forest and un-used/un-productive land. This programme has been able to impart modern skills (tapping, processing of latex etc) not known earlier to both the members of BHHs and NBHHs, it has further exposed the population living in the four selected RPCs/village to various programmes, improving their skill through various activities like imparting training and capacity development activities, encouraging participation in different programmes and meetings etc thereby contributing towards enhancement of social capital. Family members of both the BHHs and the NBHHs are associated with rubber plantation in different capacities like tapping and collection of latex, latex processing and production of rubber sheets etc which are important income generating activities leading to enhancement of financial asset/capital. Thus the introduction of the rubber plantation-based rehabilitation programme, financial asset/capital has increased consequently leading to household food security and ushering in livelihood security. Field observation suggests that during the time of field study most of the BHHs reported that they are able to provide with ease the educational and health needs of the members of their respective household after being rehabilitated by this programme. The household income has increased and the income derived from rubber plantation has made it possible for especially the BHHs to meet the extra expanses necessary to meet educational and health needs of their household members. The rehabilitation programme through rubber plantation has enhanced the socio-economic conditions of the members from both BHHs and NBHHs as this programme provides employment opportunities to the members of households not covered under this rehabilitation programme (NBHHs) as they find employment in processing of latex and smoking of rubber sheets, tapping from rubber trees and collection of latex. This has provided ample scope for income generation for the population living in such RPCs/villages.

1.7 Rubber plantation and socio-economic development

From the analysis of field data, it is observed that the BHHs of the surveyed RPCs/villages have substantially improved their socio-economic conditions in comparison to the NBHHs living in the same area. The BHHs have gained substantially in physical, financial and natural assets/capitals. The BHHs only lack in human asset/capital and slightly lack in social asset/capital in comparison to the NBHHs, this may be attributed to the fact that the BHHs are slowly giving up traditional skills and their access to *jhum* cultivation land and forests are decreasing.

It must also be mention here that the financial asset/capital has increased substantially for the BHHs as there are more BHHs which belong to higher income groups (table 5.19) in comparison to the NBHHs. The BHHs members have a continuous flow of income from the rubber plantation. The BHHs report more economically gainful activities from livestock like piggery, poultry etc. Some of them are having their own business. This has enhanced their source of income resulting to diversification of livelihood practices. However diversification of livelihood practices of the BHHs is less in comparison to the NBHHs as the NBHHs report lower income in comparison BHHs, consequently the NBHHs reports higher diversification of livelihood practices.

Hence, field investigation suggests that rubber plantation has brought about a better socio-economic development for the BHHs in comparison of the NBHHs. The lack of human asset/capital of the BHHs in comparison to the NBHHs can be made up as the members of the BHHs are in the process of learning modern skills. Access to natural asset/capital usually tends to decline with development activities but it may be mentioned here that the BHHs are emerging as individual landowners with better financial asset/capital that increases their purchasing power thereby paving in better socio-economic development.

Further reports (Bhowmik, 2006; Viswanathan and Bhowmik, 2014; DESPD, 2016) from the Government agencies like the Rubber Board, TFDPCL, TRPCL and TTAADC suggests rubber plantation in the state has been a success story. At present the TFDPCL, TRPCL and the Rubber Board are encouraging more rubber plantation in order to rehabilitate the marginalized section and improve their socio- economic conditions. The rehabilitation programme through rubber plantation scheme has enhanced the socio-economic conditions of the ST *jhumias*, landless and small farmers. In fact, the rehabilitation programme through rubber plantation is considered as one of the major development programme of the state government (Bhowmik, 2006).

During field investigation majority of the respondents were of the opinion that rehabilitation programme through rubber plantation scheme has not only played an important role in enhancing their socio-economic conditions but has also paved in proper infrastructural development. Prior to the introduction of the programme in the study area there were no proper infrastructures. But after the introduction of the programme through rubber plantation scheme by the TRPCL, many facilities such as blacktop road, health centres, new schools, water supply, tube wells, electricity, rubber processing centres etc has been setup by the TRPCL and the state government leading to the overall development of the area.

Some of the respondents of both BHHs and NBHHs of the surveyed RPCs/ villages are of the view that previously they had high incidence of poverty, some of the youth in their localities who were jobless migrated to other places for jobs, and some of them started illegal activities (minor crimes like stealing) and were vulnerable to militancy as they could be easily recruited by these militant organisation due to severe poverty, unemployment and associated depression. The respondents are of the view that the successes of rubber plantation have not only checked illegal activities but also reduced insurgency problems as it has helped in ushering in better income generating opportunities leading to peace and prosperity, thereby checking the problem of insurgency. At present a large section of the youth are employed in the rubber plantations as wage labourers or employed as processors of latex etc. With the establishment of schools and various schemes of the government to encourage literacy, the younger generation are more literate and the level of education is also increasing in comparison to the previous generations. With flow of money, the standard of living is improving but like elsewhere the old traditions are giving way to a modern attitude. Some of the respondents (mainly belonging to the older age groups) are of the view that their old tradition and culture is slowly waning off and the generation gap is increasing. Thus, the rubber plantation has slowly paved in a social change in the area, as the erstwhile *jhumias* are being brought within the ambit of modernisation bringing about livelihoods and occupational changes due to enhancement of their socio-economic conditions resulting to higher level of aspirations of especially the educated youth. This in turn open up new challenges.

1.8 Conclusion

Field findings indicate that all the five assets/capitals of the BHHs belonging to the Debbarma tribe of Dakshin Taidu, Pathaliaghat and Dakshin Dhumachhara RPCs/village have improved consequently enhancing the socio-economic conditions in comparison to the BHHs belonging to Chakma tribe of the Depachhara RPC/village. Field findings also indicates that rehabilitation programme through rubber plantation scheme involves the active participation of the local people (BHHs and NBHHs) in the four selected RPCs/villages. Besides enhancing the livelihood conditions of the BHHs, rubber plantation introduced for rehabilitation programme managed by the TRPCL has also provided employment opportunities for members from the NBHHs. This contributes immensely to livelihood sustainability and poverty reduction. Tapping rubber trees and collection of latex, latex processing and smoking of rubber sheets are main non-farm livelihood practices associated with rubber plantation. Introduction of rubber plantation has not only enhanced income generating opportunities but has also enhanced the five livelihood assets/capitals of the population living in the selected RPCs/villages of Tripura.

References

DFID. 1999. *Framework, Sustainable Livelihoods Guidance Sheets*. Retrieved from www.livelihoods.org. Accessed on 23 November 2014.

Ghosh, S., Kumar, R., Roy, C.I. and Bairy, R. 2006. Social Development Index for Social Development, in A. Kundu (ed.) *India Social Development Report, Council for Social Development,* Oxford University Press, New Delhi, pp. 219-220.

Census of India. 2011. *2011 census data-Population Enumeration Data,* Office of the Registrar General and Census Commissioner, India, New Delhi. Retrieved from http://www.censusindia.gov.in/2011census/population-enumeration.html. Accessed on 4 February, 2016.

Ghosh, S., Kumar, R., Roy, C.I. and Bairy, R. 2006. Social Development Index for Social Development, in A. Kundu (ed.) *India Social Development Report, Council for Social Development,* Oxford University Press, New Delhi, pp. 219-220.

DFID. 1999. *Framework, Sustainable Livelihoods Guidance Sheets*. Retrieved from www.livelihoods.org. Accessed on 23 November 2014.

Hasan, M.I. 2005. *Population Geography,* Rawat Publications, Jaipur, New Delhi, Bangalore, Hyderabad, Guwahati, Kolkata, pp. 40-73.

World Resources Institute (WRI) in collaboration with United Narions Development Programme, United Nations Environment Programme, and World Bank., 2005. World Resources 2005: *The Wealth of the Poor-Managing Ecosystems to Fight Poverty,* World Resources Institute, Washington, DC, pp. 3- 246.

DFID. 1999. *Framework, Sustainable Livelihoods Guidance Sheets*. Retrieved from www.livelihoods.org. Accessed on 23 November 2014.

Ghosh, S., Kumar, R., Roy, C.I. and Bairy, R. 2006. Social Development Index for Social Development, in A. Kundu (ed.) *India Social Development Report, Council for Social Development,* Oxford University Press, New Delhi, pp. 219-220.

Kamaruddin, R. and Samsudin, S. 2014. The Sustainable Livelihoods Index: A Tool to Assess the Ability and Preparedness of the Rural Poor in Receiving Entrepreneurial Project, *Journal of Social Economics Research, 1(6),* pp. 108-117.

Bhowmik, I. 2006. A status on Rubber Plantations in Tripura, *Tripura Rubber Mission- Technical Bulletin,* Tripura, pp. 1-21.

Viswanathan, P.K. and Bhowmik, I. 2014. Compatibility of Institutional Architecture for Rubber Plantation Development in North East India from the Comparative Perspective of Kerala, *NRPPD Discussion Paper No. 38,* Centre For Development Studies, Thiruvananthapuram, pp. 1-71.

DESPD. 2016. *Economic Review of Tripura (2015-16), 17th issue,* Government of

Tripura, Agartala, Tripura, pp. 3-151.

Bhowmik, I. 2006. A status on Rubber Plantations in Tripura, *Tripura Rubber Mission- Technical Bulletin,* Tripura, pp. 1-21.

6
Summary of Findings and Conclusions

Introduction

The state of Tripura was a princely state during the colonial period whose territory covered portions lying in present Bangladesh. The present state of Tripura was dominated by the *Borok* group of people speaking *kokborok* language. However with partition of India in 1947 followed by Bangladesh Liberation War, the demographic composition of the state has changed, at present the indigenous tribes has been outnumbered by the immigrant population from adjoining Bangladesh. Due to the location, the state experiences immigration resulting to population pressure this in turn has brought about problems of land alienation and marginalization in jobs, professions and politics faced by especially the ST population of the state whose exposure to modernity is recent in comparison to the plains people.

The topography of Tripura consists of hillocks (locally termed as *tilla*) and fertile valleys (locally termed as *lunga*). Most of the ST inhabits the rural remote hilly tracts of the state where livelihoods are dependent on the surrounding environment, here forest resources are important besides primitive subsistence type of agriculture, *jhum* cultivation is prevalent. The ST population are relatively less aware about modern farming techniques in comparison to the immigrant population who mainly occupy the fertile valleys or *lunga* and practice settled agriculture.

With the merger of the princely state of Tripura with the India Union in 1949, and the formation of full-fledged state of Tripura in 1972, the state administration has initiated various tasks to tackle the problems for rehabilitating millions of refugees from Bangladesh (erstwhile East Pakistan), resettlement of *jhumias* etc however did not get its due attention in the initial phase of development programme of the state. Despite the initiation of various schemes by the state government for restoration of land to the ST population, thousands of ST population are still landless.

Due to the remote location surrounded on three sides by Bangladesh and connected to India only through a small stretch of land through the northern-east part of Cachar district of Assam (this too created by the partition of the country as Tripura lost its easy and direct connectivity to Kolkata as this route passes through present Bangladesh). The state lacks in infrastructure development as connectivity is restricted hence there is lack of diversification in economic activities. Population pressure coupled with low economic development has resulted in high incidence of unemployment in the state. However since the past two to three decades, the state of Tripura is changing and has emerged as the second largest producer of NR in the country.

The present thesis highlights on the growth and distribution of rubber plantation in Tripura and its impact on the socio-economic development and livelihood diversification of the people associated with rubber plantation.

Rubber plantation was introduced by the government here around 1963 on a trial basis as a soil conservation measure on the degraded forests prone due to large scale felling of tress for commercial purpose coupled with the practice of *jhum* cultivation. As the trial plantation of rubber trees has been a success, it has encouraged further expansion of rubber plantation on the degraded lands; hence the state agencies came up with the innovative idea to rehabilitate the marginalized section of population mainly *jhumias,* landless and small farmers through rubber plantations and slowly coax them into the ambit of settled agriculture practice, thereby checking *jhum* cultivation and associated soil erosion and deforestation, further with rubber plantation this marginalized section can be provided with diversification and alternative livelihood practices, enhancing their socio-economic conditions, check further deforestation and soil erosion as well as promote production of NR thereby enhancing the development of the state. Consequently, the focus of this research as already mentioned are to assess the nature of expansion of rubber plantation, its distribution over the state, the role of different agencies in the promotion and development of rubber plantation in the state. Further the impact of rubber plantation-based rehabilitation programme has also been looked into in order to examine whether rubber plantation-based rehabilitation programme has really achieved its goal in providing the marginalized section of peoples with alternative and diverse livelihood opportunities bringing about an improvement in their socio- economic conditions.

The specific objectives placed before the research has been- (i) to examine the spatial pattern of growth and distribution of rubber plantation in the state, (ii) to assess the role of different agencies on the growth and distribution of rubber plantation in the state and (iii) to evaluate the contributions of rubber plantation on the socio-economic development of the marginalized ST families in providing them with alternative livelihood options and diversification of livelihoods.

This research is divided into six chapters which consist of two broad sections: one section is totally based on discussion and analysis of secondary data and another is based on discussion and analysis of primary data generated through field survey.

From secondary data care has been taken to assess the expansion of area under rubber plantation by the help of time series analysis, nature of gain, rate of annual growth in area under rubber plantation and its distribution in the state and the spatial distribution of area under rubber plantation in the state has also been measured and discussed. Further an attempt has been made to assess the role of different agencies on the growth and distribution of rubber plantation in the state. For computing the growth rate of area under rubber plantation in the state the 'arithmetic growth rate' has been used. 'three yearly moving average' has been used for computing the trend line which suggest the nature of annual gain in area under rubber plantation in the state. Concentration index has been used for computing the district wise distributions of area under rubber plantation in the state. RS and GIS technique using ERDAS IMAGINE 9.1 and ArcGIS 10.3 software has been used for analyzing the spatial distribution of area rubber plantation in the state.

From primary data an attempt has been made to assess the impact of rubber plantation-based rehabilitation programme. Primary data has been collected through field investigation in four selected rubber plantation centre (RPC) or villages where TRPCL has developed rubber plantation. TRPCL has been selected as TRPCL main objective is to economically rehabilitate the ST *jhumias,* landless and small farmers principally through rubber plantation, further it is the only organisation which has its presence in all the four districts of the state as TRPCL managed RPCs are found in all the four districts, TRPCL also reports the highest number of RPCs in comparison to others agencies looking after rubber plantation in the state. The TRPCL has 68 RPCs distributed in all the four districts.

For selection of the RPCs where field survey has been conducted, the RPCs managed by TRPCL has been arranged as per district, this has been further categorized block wise. From the blocks having the maximum RPCs, one RPC from each district has been selected at random for field investigation. The four selected RPCs/villages are: Depacherra RPC/village located in Pencharthal block of North Tripura district, Dakshin Dhumacherra RPC/village located in Manu block of Dhalai district, Pathaliaghat RPC/village located in Bishalgarh block of West Tripura district and Dakshin Taidu RPC/village located in Ampi block of South Tripura district.

Primary data has been collected at two levels- (i) through focused group discussion (FGD) with the village elders and concerned officials (TRPCL) of the selected RPC/village where information on the general background of the RPC/ village, regarding the rubber plantation rehabilitation programme etc has been collected, and (ii) through household survey where data on the demographic and socio-economic conditions of the household, livelihood practices, indicators suggesting the livelihood assets associated with the five livelihood assets/capitals viz. physical, natural, human, social and financial assets/capitals of the sustainable livelihood approach (SLA) has been collected with the help of semi-structured schedule by interviewing two sets of households i.e. 1st set comprising of beneficiaries households (BHHs) who are rehabilitated through rubber plantation and the 2nd set comprising of households not practicing rubber plantation (NBHHs) but belonging to the same community and living in close proximity to each other, so to measure and compare the livelihood assets and diversification of livelihoods of the BHHs with that of the NBHHs.

The modified version of Simpsons Index of diversity (SID) has been used for measuring and comparing the livelihood diversity of the BHHs with that of the NBHHs. Range Equalization Method (UNDP method of human development index) has been used in order to compute the composite score of the five livelihood assets/capitals for measuring and comparing the five livelihood assets/capitals of the BHHs with that of the NBHHs. The composite index score of the five livelihood assets/capitals of both the BHHs and NBHHs has been diagrammatically represented by livelihood asset pentagon in order to know the strength and weakness of all the five assets/capitals of both the sets of surveyed households. Major Findings.

The annual share of area under rubber plantation of Tripura to the total area under rubber plantation of India is increasing continuously. It is found that during 1972-73 the share was only 0.05 percent which has increased to the tune of 9.17 percent during 2015-16. From 1972-73 to 2015-16, the share of area under rubber plantation of the state to the total area under rubber plantation of India has been increasing at an average rate of 3.88 percent per annum.

From 1972-73 (103 ha) to 2015-16 (74,335 ha) there has been an expansion of 74,232 ha of area under rubber plantation in the state with an overall growth rate as high as 72,069.90 percent within a span of about five decades. Consequently, of the 1 lakh ha of the state's geographical area found suitable for rubber plantation, 74,335 ha (74.34 percent) of area has already come under rubber plantation by 2015-16 suggesting that around 25,665 ha (25.66 percent) of area earmarked for rubber is still available for further expansion.

It is observed that total area under rubber plantation in the state (1972-73 to 2015-16) is continuously increasing annually but at present at a declining growth rate with fluctuating trend. The highest growth rate in area under rubber plantation in the state was experienced in the initial stage, during 1973-74 as that year reported more than 100 percent annual growth rate (116.50 percent) while the lowest annual growth rate has been observed in the year 2013-14 with just 2.12 percent. From 1972-73 to 2015-16, area under rubber plantation in the state has been increasing at an average annual growth rate of 18.05 percent per annum.

The annual gain in area under rubber plantation in the state is found to be lowest during 1974-75 with a marginal gain of 84 ha while the highest gain in area under rubber plantation has been observed during 2014-15 (7,767 ha). From 1973-74 to 2015-16, annual gain in area under rubber plantation of the state has been at an average rate of 1,726.33 ha per annum. Findings suggest that there is a large variation in the annual gain in area under rubber plantation in the state as the coefficient of variation is to the tune of 98.92.

The trend in the annual gain in area under rubber plantation in the state suggest that area under rubber plantation has been slowly increasing from 1974-75 to 1984-85. After 1985, the gain in area under rubber plantation started increasing at a faster pace reaching its peak in 1987-88. After 1988, the gain in area under rubber plantation suggests a declining trend till 1992-93; however, after 1993 the gain in area under rubber plantation peaked up again, this continued till 1999.

After 2000, the gain in area under rubber plantation is suggesting a fluctuating trend till 2005-06. After 2006, the gain in area under rubber plantation again sharply increased reaching a third peak in 2008-09. After the year 2009 the gain in area under rubber plantation started declining drastically which continued till the year 2012-13. From the year 2014, the gain in area under rubber plantation in the state is peaking up again.

From the year 2008-09 to 2015-16, the share of area under rubber plantation is highest in West Tripura district followed by South Tripura district, North Tripura district and the least area under rubber plantation is in Dhalai district.

Area under rubber plantation has been continuously increasing from 2008-09 to 2015-16 in all the districts. In case of the share of area under rubber plantation in each district to the total area under rubber plantation from 2008-09 to 2015-16, it is observed that the share of area under rubber in West Tripura, North Tripura and Dhalai districts are continuously increasing, however in Dhalai district the share of area under rubber plantation is on the decline in the recent years, the share in area under rubber plantation in South Tripura district suggests a declining trend except in 2010-11 where the share is found marginally higher (by 0.29 %) than the previous year i.e. 2009-10.

The growth rate in area under rubber plantation in all the four districts from 2009-10 to 2015-16 suggest a fluctuating trend. The highest growth rate of area under rubber plantation is noticed during the year 2014-15 in North Tripura district (25.79 %), followed by West Tripura district (18.24 %) and South Tripura district (8.35 %). It is interesting to note that Dhalai district of Tripura reports negative growth rate in area under rubber plantation for the year 2014-15.

During the year of 2015-16, the highest concentration of area under rubber plantation in the state is found in West Tripura district with a concentration index (CI) score of 1.77. South Tripura district has slightly lower concentration of area under rubber plantation with a CI score of 1.06. North Tripura district has low concentration of area under rubber plantation with a CI score of 0.68 while Dhalai district has very low concentration of area under rubber plantation with a CI score as low as 0.25.

The spatial distribution of area under rubber plantation in the state computed from two landsat images (2001 to 2015) suggest that area under rubber plantation is spreading, but concentrated mostly towards the western and southern parts of the state as the topography here is ideal for rubber plantation marked by *tillas* as the slopes of the *tillas* are ideal for rubber plantation.

Finding suggests that the TFDPCL, TRPCL and the Rubber Board have played an important role in the expansion of area under rubber plantation in the state. TFDPCL has contributed the maximum in expansion of area under rubber plantation as it has developed 11,624 ha (15.64 percent) of rubber plantation followed by the TRPCL with 9,154 ha (12.31 percent) of rubber plantation while Rubber Board has developed about 3,775 ha (5.08 percent) of rubber plantation in the state. It is observed that the share of area under rubber plantation developed by the three agencies to the total area under rubber plantation of the state is declining at present as private owners and individual rubber planters are emerging.

Findings also suggest that the annual growth of area under rubber plantation managed by the three agencies are found to be increasing but at a declining growth rate with mild fluctuating trend.

It is observed that till the year 2014-15, TRPCL was the only agency that developed rubber plantation in all the four districts of the state i.e. North Tripura, Dhalai, West Tripura and South Tripura districts while the presence of TFDPCL and the Rubber Board are restricted to only three districts (North Tripura, West Tripura and South Tripura) and are totally absent in Dhalai district of the state.

As per data of 2014-15, TFDPCL has developed maximum area under rubber plantation in South Tripura district (44.48 %) followed by West Tripura district (30.88%) and North Tripura district (24.64 %). TRPCL has developed about 41.12 percent of area under rubber plantation in West Tripura district which is the highest followed by South Tripura district (23.48 %), North Tripura district (18.02 %) and Dhalai district (17.38 %). In case of Rubber Board it has developed maximum area under rubber plantation in West Tripura district (77.62 %) followed by South Tripura district (16.74 %) and North Tripura district (5.64 %). These suggest that both the TRPCL and the Rubber Board have maximum area under rubber plantation in West Tripura district while TFDPCL have maximum area under rubber plantation in South Tripura district of Tripura.

In case of district wise contribution of area under rubber plantation developed by the TFDPCL, TRPCL and Rubber Board to the district wise total area under rubber plantation of the state during 2014-15, TFDPCL has contribution the highest area under rubber plantation in North Tripura district (29.49 %) followed by South Tripura district (22.94 %) and the least contribution is in West Tripura district with 9.80 percent. TRPCL has maximum contribution of area under rubber plantation in Dhalai district (39.39 %) followed by North Tripura, West Tripura and South Tripura districts with 16.89 percent, 10.22 percent and 9.48 percent respectively. Rubber Board on the other hand has large contribution of area under rubber plantation in West Tripura district (8.30 %) followed by South Tripura and North Tripura districts with 2.91 percent and 2.27 percent respectively.

The analysis of field data suggests that from the four RPCs/villages surveyed; households are mainly belonging to three tribes i.e. Debbarma, Reang and Chakma tribes who have been rehabilitated through rubber plantation scheme managed by the TRPCL. Household belonging to Debbarma tribe are found in two RPCs/villages i.e. Pathaliaghat and Dakshin Taidu RPCs/villages, Chakma tribe is concentrated in one RPC/village i.e Depachhara RPC/village while in Dakshin Dhumachhara reports mixed population compo\ition where 76 percent of the household are belonging to Debbarma tribe and about 24 percent of the surveyed BHHs belong to Reang tribe. Hence field data suggest that the Debbarma tribe has a higher share as beneficiaries of this rehabilitation programme.

Field finding suggest that in all the four RPCs/villages, the physical asset/capital are found to be higher in comparison to other four assets/capitals. Human asset/capital of the BHHs of the four RPCs/villages suggest that the BHHs from Pathaliaghat RPC/village are better off in comparison to the other three RPCs/

villages as the composite index score of the human asset/capital of Pathaliaghat RPC/village is as high as 0.43, closely followed by Dakshin Dhumachhara RPC/village (0.38 score) and Dakshin Taidu RPC/village (0.37 score). The Depachhara RPC/village has the least human asset/capital composite index score of 0.29. High literacy rate (83.90 %) and modern skills known (100 %) by the members of the household of the Pathaliaghat RPC/village have attributed for the high human asset/capital among the BHHs of the Pathaliaghat RPCs. Mention must be made that the Debbarma tribe constitute 100 percent BHHs of the Pathaliaghat RPC.

In regard to social asset/capital of the BHHs of the four RPCs/villages, Dakshin Dhumachhara RPC/village has the highest score (0.54), this is closely followed by Dakshin Taidu RPC/village (0.51 score) which is slightly higher than the composite index score of Pathaliaghat RPC/village which has a score of 0.50 while Depachhara RPC/village has the least social asset/capital composite index score (0.47). BHHs from Dakshin Dhumachhara RPC/village reports having higher participation in programme or meeting organized by rubber growing agencies and others government and non-government organization. Social asset/capital suggest that BHHs belonging to Debbarma tribe of the Dakshin Dhumachhara, Dakshin Taidu and Pathaliaghat RPCs/villages have higher social asset/capital in comparison to the BHHs belonging to Chakma tribe living in Depcahhara RPC/village.

It is interesting to note that in case of natural asset/capital the BHHs of Dakshin Dhumachhara RPC/village score slightly higher (0.46) than the other RPCs/villages. Next comes Dakshin Taidu with a composite index score of 0.44 followed by Depachhara RPC/village (0.38 score) and Pathaliaghat RPC/village with a composite index score of 0.37 (appendix 12). High access to landholding size between 1 – 1.5 ha (80.00 %) and access to water sources like Pond/Stream/River (60.00 %) by the BHHs of Dakshin Dhumachhara RPC/village are the main attributing factors for high natural asset/capital of the BHHs living in Dhumachhara RPC/village than other RPCs/villages. With a composite index score of 0.46, the natural asset/capital of Dakshin Dhumachhara RPC/village is the highest, this is closely followed by Dakshin Taidu with a score of 0.44, followed by Depacherra RPC/village (0.38 score) and Pathaliaghat RPC/village i.e. 0.37. The Debbarma tribe of Dakshin Dhumachhara and Dakshin Tiaidu RPCs/villages reports having better access to natural asset/capital than the Chakma tribe of Depachhara RPC/village.

In regard to financial asset/capital, the BHHs of Dakshin Taidu RPC/village are found to be better off than other RPCs/villages. Dakshin Taidu RPC/village has a score of 0.45 in financial asset, followed by Dakshin Dhumachhara RPC/village, Dakshin Dhumachhara RPC/village, Pathaliaghat RPC/village. Depachhara RPC/village reports the least score in financial asset/capital i.e. 0.31. Higher share of BHHs in high income group are living in Dakshin Taidu RPC/village. Thus this becomes the main attributing factor for high financial asset/capital of the BHHs living here in comparison to other RPCs/villages. About 33 percent of the BHHs of Dakshin Taidu RPC/village earn income ranging between Rs 9,001 – 12,000 per month, 18.18 percent of the surveyed BHHs earn an income ranging between Rs 12,001 – 15,000 per month and 1.82 percent of the BHHs of Dakshin Taidu RPC/village earn income ranging between Rs 15,001 – 18,000 per month (appendix

14). In regard to financial asset/capital, the BHHs belonging to Debbarma tribe of Dakshin Tiaidu, Dakshin Dhumacherra and Pathaliaghat RPCs/villages are found to have higher financial asset/capital than the Chakma tribe of Depacherra RPC/ village.

The analysis of field data suggest that both the BHHs as well as the NBHHs are found engaged in various livelihood activities associated with rubber plantation. Field data suggest that about 97.14 percent of the BHHs are practicing tapping of rubber plantation as their major livelihood practice. 33.66 percent and 13.86 percent of the respondents from the NBHHs report that their major livelihood practices are as tappers of rubber plantation and processing of latex respectively. 1 percent of the NBHHs is employed in rubber processing centre as guard and is thereby a livelihood directly related with rubber plantation. Other major livelihood practices engaged by both the BHHs and NBHHs are shop keeping, transport business mainly engaged as drivers, carpentry business, petty business in local market, mechanic and party decorators.

Field data suggests that the NBHHs have relatively high diversification of livelihood practices (SID value of 0.68) when compared to the BHHs (SID value of 0.32) as rubber plantation gives relatively higher and steady income while the NBHHs having lesser income and needs to diversify their source of income consequently diversification of livelihood practices is higher among the NBHHs in comparison to the BHHs.

Field finding suggest that BHHs are better off in physical asset/capital, natural asset/capital and financial asset/capital in comparison to the NBHHs. The composite index score of the five assets/capitals associated with livelihood pentagon suggests that the BHHs have gained substantially in physical asset/ capital (0.83 score), natural asset/capital (0.43 score) and financial asset/capital (0.36 score) in comparison to the NBHHs as the NBHHs report a composite index score of 0.68, 0.40 and 0.29 score respectively. The BHHs slightly lack in social asset/capital (0.50 score) and human asset/capital (0.33 score) in comparison to the NBHHs with composite index score of 0.52 and 0.48 score respectively.

Among the five assets/capitals, field finding suggest that only physical asset/ capital of the BHHs are found to be highly sustainable while other four assets/ capitals i.e. social, natural, financial and human assets/capitals are found to be moderately sustainable. On the other hand, among the five assets/capitals of the NBHHs, physical and social assets/capitals are found to be moderately high sustainable while human, natural and financial assets/capitals are found to be moderately sustainable.

Field investigation suggests that rubber plantation has not only enhanced the socio-economic conditions but has also paved in proper infrastructural development. Prior to the introduction of rubber plantation in the study area there were no proper infrastructures. But after the introduction of rubber plantation in the area by the TRPCL, the state government along with the different agencies, has provided many facilities such as all weather black top road, health centres, schools, water supply, tube wells, electricity, rubber processing centres etc has developed leading to the overall development of the area. Mention must be made here that besides the rubber plantation-based rehabilitation programme the state and the central governments like elsewhere in the country has introduced here various programmes for the upliftment of the marginalized. Both the BHHs and NBHHs reported that previously they had high incidence of poverty resulting to substantial migration of the working age group to other areas in search of employment. Some of the youth in the surveyed localities took recourse to petty criminal offences like theft etc due to poverty, unemployment, and frustration. Some of the respondents are of the view that the successes of rubber plantation have checked insurgency problems as it has helped in ushering in peace and prosperity in their area which in the past was a hotbed of insurgency. However, at present a large section of the youth are employed in the rubber plantation as wage labourers or employed as latex processors etc. With flow of money, the standard of living here has improved but like elsewhere the old traditions are giving way to a modern attitude. Some of the respondents (mainly belonging to the older age groups) are of the view that their old tradition and culture is slowly waning off and the generation gap is increasing.

Field findings suggest that the rehabilitation programme through rubber plantation has enhanced the socio-economic conditions of the members from both BHHs and NBHHs as this programme provides employment opportunities to the members of households not covered under this rehabilitation programme (NBHHs) as they find employment in processing of latex and smoking of rubber sheets, tapping from rubber trees and collection of latex and as wage labourers. This has provided ample scope for income generation for the population living in the surveyed RPCs/villages. It can be safely inferred that a similar condition prevails in the other RPCs not covered by field investigation.

Conclusion

It can be concluded that the rehabilitation programme through rubber plantation scheme in the state is a success and the area under has been increasing continuously which in recent years is increasing but at a declining growth rate. Rubber plantation has spread throughout the state but has highest concentration in West Tripura district followed by South Tripura district, North Tripura and Dhalai districts.

Three agencies viz. TFDPCL, TRPCL and the Rubber Board are playing an important role in the expansion of area under rubber plantation in the state. TFDPCL has contributed the maximum towards the expansion of area under rubber plantation followed by TRPCL and the Rubber Board.

Rubber plantation in Tripura has changed the *jhumias* and landless families (now BHHs) into land owners, paving in a social change as reflected by livelihood practices, there has also been a shift of livelihood practices of both the BHHs and NBHHs. At present both the groups are engaged in various livelihood activities associated with rubber plantation and have taken up different types of business. This has not only generated flow of income from rubber plantation but also from other sources, consequently enhancing their socio-economic conditions. However, there has been a disparity in the socio-economic development among the different tribes rehabilitated through rubber plantation scheme. Field data suggest that out of the five livelihood assets/capitals the Debbarma tribe are found to score higher in most of the livelihood asset in comparison to the Chakma and Reang tribes.

Field finding also suggests that in spite of the fact that the rehabilitation through rubber plantation is more or less having a positive impact on the socio-economic development yet there are certain problems. These problems are associated with lack of protective measures during latex collection and processing the rubber sheets, no fixed payment as NR price fluctuate as per the national and international price market. Thus the TRPCL needs to provide the tappers with modern protective gears during collection of latex as well as during processing of latex. The agency also needs to look into the fluctuating price of the NR so that the rehabilitated households can maintain their flow of income for sustaining their livelihoods.

Bibliography

Addisu, Y. 2017. Livelihood strategies and diversification in western tip pastoral areas of Ethiopia, *Pastoralism: Research, Policy and Practice, 7:9,* pp. 1-9.

Agyeman, B.A.S., Brempong, S.A. and Onumah, E.E. 2014. Determinants of Income Diversification of Farm Households in the Western Region of Ghana, *Quarterly Journal of International Agriculture, Vol. 53, No. 1,* pp. 55-72.

Ahmed, M.T., Bhandari, H., Gordoncillo, P.U., Quicoy, C.B. and Carnaje, G.P. 2015. Diversification of rural livelihoods in Bangladesh, *Journal of Agricultural Economics and Rural Development, Vol. 2(2),* pp. 032-038.

Ahrends, A., Hollingsworth, P.M., Ziegler, A.D., Fox, J.M., Chen, H., Su, Y. and Xu, J. 2015. Current trends of rubber plantation expansion may threaten biodiversity and livelihoods, *Global Environment Change, 34,* pp. 48-58.

Anandan, G., Thomas, A., Benickson, C., Chitra, DR., Geethu. M., et al. 2014. Estimation of Tree Species Diversity in Four Campuses of Roever Institutions using Simpsons Diversity Index, *Journal of Biodivers Endanger Species. Volume 2, Issue 4,* pp. 135.

Army Map Service. 1955. Toposheet No. NF 46-1, NF 46-2, NF 46-6, NG 46-13 and NG 46-24, Series U502, Scale 1: 250,000, Corps of Engineers, U.S.Army, Washington, D.C.

Bareh, H.M. 2001. *Encyclopaedia of North-East India-Volume viii-Tripura,* Mittal Publications, New Delhi, pp. 1-279.

Bera, G.K. 2009. Introduction, in G.K. Bera, B.K. Medhi, R.P. Athparia, and K. Jose (ed.) *Tribal Development in Tripura,* EBH Publisher, Guwahati, pp. 1-16.

Bhattacharjee, S.R. 1989. *Tribal Insurgency in Tripura: A Study in Exploration Causes,* Inter- India Publications, New Delhi, pp. 23-28.

Bhattacharjee, S 2012. Multidisciplinary research can make the medicinal plant industry in Tripura viable- A conceptual study based on tows matrix, *International Journal of Multidisciplinary Research, Vol 2, Issue 5,* pp. 146- 154.

Bhattacherjee, P.N. and Singh, R.G. 1995. *Tea Plantation and the Tribes of Tripura,* Tripura State Tribal-Cultural Research Institute and Museum, Government of Tripura, pp. 1-88.

Bhattacharya, B.K. 1980. Pattern of Administrative Organisation, in J.C. Chaudhuri (ed.) *Tripura- The Land and its People,* Leeladevi Publication, Delhi, pp. 44- 54.

Bhattacharya, S. 1992. *From Jhuming to Tapping,* Directorate of Research, Department of Welfare for Schedule tribes, Government of Tripura, Agartala, pp. 1-85.

Bhattacharyya, T., Sarkar, D., Gangopadhyay, S.K., Dubey, P.N., Baruah, U., Chamuah, G.S., Mukhopadhyay, S., Nayak, D.C., Maji, A.K., Saxena, R.K., Barthwal, A.K., Krishna, N.D.R., Mandal, C. and Sahgal, J. 1998. Soils of Tripura.II- Suitability for Rubber, *Agropedology, 8,* pp. 55-60.

Bhatt, S.C. and Bhargava, G.K. 2006. *Land and People of Indian state and Union territories (In 36 Volumes)- Tripura, Volume- 26,* Kalpaz Publications, Delhi, pp. 17-102.

Bhaumik, S. 2015. Tripura: Counter-insurgency and Governance, in R. Samaddar (ed.) *Government of Peace: Social Governance, Security and the Problematic of Peace,* Ashgate Publishing Ltd., Wey Court East, England, pp. 79-106.

Bhowmik, I. 2006. A status on Rubber Plantations in Tripura, *Tripura Rubber Mission- Technical Bulletin,* Tripura, pp. 1-21.

Bhowmik, I. and Viswanathan, P.K. 2015. Emerging Labour Relations in the Small Rubber Plantations of Tripura, *NRPPD, Discussion Paper No. 47,* Centre For Development Studies, Thiruvananthapuram, pp. 1-33.

Bordaloi, B.N. 1990. The Problem of Rehabilitation of the Jhumiya Families in the Hills Areas of Assam, in D.N. Majumdar (ed.) *Shifting Cultivation in Northeast India,* Omsons Publication, New Delhi, pp. 226-228.

Borthakur, D.N. 1992. *Agriculture of the North eastern region: with special references to hill Agriculture,* BeeceePrakashan, Guwahati, pp. 59-71.

Burma, K.K. 1990. Shifting Cultivation in the Hills of Assam, in D.N. Majumdar, (ed.) *Shifting Cultivation in Northeast India,* Omsons Publication, New Delhi, pp. 226-228.

Census of India. 2011. *2011 census data-Population Enumeration Data,* Office of the Registrar General and Census Commissioner, India, New Delhi. Retrieved from http://www.censusindia.gov.in/2011census/population_enumeration.html. Accessed 4 February 2016.

Centre for Development Studies. *Area under rubber and average yield per hectare.* Retrieved from http://cds.edu/research/ru/nrppd/database-on- plantation/rubber/. Accessed on 23 January, 2016.

Chakravarti, M. 1986. Forestry in Tripura: Its Past and Present, in M.D. Gupta, A.K. Gangopadhyay, T. Bhattacharya, M. Chakraborti (ed.) *Forestry Development in North-East India,* Omsons Publications, Guwahati and New Delhi, pp. 31- 41.

Chatterjee, S.N. 1984. *Tripura- A Profile,* Inter-India Publications, New Delhi, pp. 1- 61.

Chattopadhyay, S. 2015. Environmental Consequences of Rubber Plantations in Kerala, *NRPDD, Discussion Paper No. 44*, Centre for Development Studies, Thiruvananthapuram, pp. 1-53.

Chaudhuri, N.R.R. 1980. The Historical Past, in J.C. Chaudhuri (ed.) *Tripura- The Land and its People*, Leeladevi Publication, Delhi, pp. 13-43.

Chaudhuri, P.S., Bhattacharjee, S., Dey, A., Chattopadhyay, S., and Bhattacharya, D. 2013. Impact of age of rubber (Hevea brasiliensis) plantation on earthworm communities of West Tripura (India), *Journal of Environmental Biology, Vol. 34*, pp. 59-65.

Choudhury, J. 2009. Efforts of Tribal Development in Tripura: An Assessment, , in

G.K. Bera, B.K. Medhi, R.P. Athparia, and K. Jose (ed.) *Tribal Development in Tripura*, EBH Publisher, Guwahati, pp. 121-135.

Chen, H., Yi, Z-F., Schmidt-Vogt, D., Ahrends, A., Beckschafer, P., Kleinn, C., Ranjitkar S. and Xu, J. 2016. Pushing the Limits: The Pattern and Dynamics of Rubber Monoculture Expansion in Xishuangbanna, SW China, *PloS ONE 11(2)*, pp. 1-15.

Cobley, L.S. 1976. *An Introduction to the Botany of Tropical Crops*, Longman Group Limited London, pp. 336-344.

Cochran, W. G. 1977. *Sampling techniques (3rd ed.)*, New York: John Wiley & Sons.

Dasgupta, M. 1986. Jhumias of Tripura, *Economic and Political Weekly, Vol. 21, No.*

44/45, pp. 1955-1957, 1959-1960.

Das, R. 2014. Rubber Plantation: Tripura Scenario, in K. Shib (ed.) *Rubber in Tripura (1963-2013)- Golden Jubilee Celebration, Souvenir*, Bina Printers, Arundhatinagar, Agartala, pp. 25-33.

Das, S., Choudhury, S. and Roy, A. 2012. The Success Story of Rehabilitation of Jhumias in Tripura- A study on Baramura-Deutamura Range, *International Journal of Engineering and Science, Vol. 1, Issue 10*, pp. 25-29.

Debbarma, B. 2014. Tripura Rehabilitation Plantation Corporation Limiited, in K. Shib (ed.) *Rubber in Tripura (1963-2013)- Golden Jubilee Celebration, Souvenir*, Bina Printers, Arundhatinagar, Agartala, pp. 35-42.

Debbarma, R. 2013. *Socio-Economic Impact Of Rubber Plantation On Shifting Cultivators In Dhalai District Of Tripura*, M.Phil Dissertation, North-Eastern Hill University, Shillong, Meghalaya, pp. 54-135.

Debbarma, S. 2009. Refugee Rehabilitation and Land Alienation in Tripura, in W. Fernandes and S. Barbora (ed.) *Land, People and Politics: Contest over Tribal Land in Northeast India*, North-Eastern Social Research Centre, Guwahati and International Work Group for Indigenous Affiars (IWGIA), Denmark, pp. 113-127.

Debele, B.N. and Desta, G.D. 2016. Livelihood Diversification: Strategy, Determinants and Chaallenges for Pastoral and Agro-Pastoral Communities of Bale Zone, Ethiopia, *International Review of Social Sciences and Humanities, Vol. 11, No. 2*, pp. 37-51.

DESPD. 2008. *Economic Review of Tripura (2007-08), 9th issue,* Government of Tripura, Agartala, Tripura, pp. 3-153.

DESPD. 2009. *Economic Review of Tripura (2008-09), 10th issue,* Government of Tripura, Agartala, Tripura, pp. 3-142.

DESPD. 2010. *Economic Review of Tripura (2009-10), 11th issue,* Government of Tripura, Agartala, Tripura, pp. 3-131.

DESPD. 2011. *Economic Review of Tripura (2010-11), 12th issue,* Government of Tripura, Agartala, Tripura, pp. 3-151.

DESPD. 2012. *Economic Review of Tripura (2011-12), 13th issue,* Government of Tripura, Agartala, Tripura, pp. 3-46.

DESPD. 2013. *Economic Review of Tripura (2012-13), 14th issue,* Government of Tripura, Agartala, Tripura, pp. 3-152.

DESPD. 2014. *Economic Review of Tripura (2013-14), 15th issue,* Government of Tripura, Agartala, Tripura, pp. 3-157.

DESPD. 2015. *Economic Review of Tripura (2014-15), 16th issue,* Government of Tripura, Agartala, Tripura, pp. 144-157.

DESPD. 2016. *Economic Review of Tripura (2015-16), 17th issue,* Government of Tripura, Agartala, Tripura, pp. 90-104.

Dev, T., Sultana, N. and Hossain, M.E. 2016. Analysis of the Impact of Income Diversification Strategies on Food Security Status of Rural Households in Bangladesh: A Case Study of Rajshahi District, *American Journal of Theoretical and Applied Business, Vol. 2, No. 4,* pp. 46-56.

Dev Varman, S.B.K. 1999. A Study Over The Jhum And Jhumia Rehabilitation In The Union Territory Of Tripura, *Directorate of Research, Department of Welfare for Scheduled Tribes and Scheduled Castes, Government of Tripura, Agartala,* Print Best, Agartala, Tripura, pp. 9-39.

Devi, A.I. 2010. *Amazing North East - Tripura,* Vij Books India Pvt. Ltd. New Delhi, pp. 75-77.

DFID. 1999. *Framework, Sustainable Livelihoods Guidance Sheets.* Retrieved from *www.livelihoods.org*. Accessed on 23 November 2014.

Dong J. W., Xiao X. M., Chen B. Q., Torbick, N., Jin, C., Zhang, G. and Biradar, C. 2013. Mapping deciduous rubber plantations through integration of PALSAR and multi-temporal Landsat imagery, *Remote Sensing of Environment,* vol. 134, pp. 392-402.

Fabusoro, E., Omotayo, A.M., Apantaku, S.O. and Okuneye, P.A. 2010. Forms and Determinants of Rural Livelihoods Diversification in Ogun State, Nigeria, *Journal of Sustainable Agriculture,* 34:4, pp. 417-438.

FAO. 2013. Food and Agriculture Organization of the United Nations, *Processing of Natural Rubber.* Retrieved from http://ecoport.org/ep?SearchType=earticleView&earticleId=644&page=4343. Accessed on 4 February 2016 and on 10th March 2017.

Fox, J. and Castella, J.C. 2013. Expansion of rubber (Hevea brasiliensis) in Mainland Southeast Asia: What are the prospects for smallholders?, *Journal of Peasant Studies, 40 (1)*, pp. 155-170.

Fu, Y., Brookfield, H., Guo, H., Chen, J., Chen, A. and Cui, J. 2009. Smallholder rubber plantation expansion and its impact on local livelihoods, land use and agrobiodiversity, a case study from Daka, Xishuangbanna, southwestern China, *International Journal of Sustainable Development and World Ecology, Vol. 16, No.1*, pp. 22-29.

Gan-Chaudhuri, J. 1980. The Perennially Charming Tripura, in J. Gan-Chaudhuri (ed) *Tripura: The Land and its People*, Leeladevi Publications, Delhi, pp. 1- 12.

Geosrge, J.G. and Chandrashekar, H.M. 2014. Growth and Trends in Production and Marketing of Natural Rubber in Kerala, India, *International Journal of Current Research and Academic Review, 2(8)*, pp. 53-61.

George K. T., Haridasan, V. and Sreekumar, B. 1988. Role of Government and Structural Changes in Rubber Plantation Industry, *Economic and Political Weekly,Vol. 23, No. 48*, pp. M158-M166.

Ghosh, S., Kumar, R., Roy, C.I. and Bairy, R. 2006. Social Development Index for Social Development, in A. Kundu (ed.) *India Social Development Report, Council for Social Development*, Oxford University Press, New Delhi, pp. 219-220.

Gleason, C. 2005. *The Biography of Rubber*, Crabtree Publishing Company, pp. 1-32. Government of India. 2003. Successful Governance Initiatives and Best Practices: Experiences from Indian States, *Planning Commission (Government of India), Human Development Resource Centre, UNDP*, Academic Foundation, New Delhi, pp. 50-54.

Government of Tripura (GoT). 2007. *Tripura Human Development Report*, Government of Tripura, Tulika Print Communication Services, New Delhi, pp. 1-32.

Gupta, A.K. 2000. Shifting Cultivation and Conservation of Biological Diversity in Tripura, Northeast India, *Human Ecology, Vol. 28, No. 4*, pp. 605-629.

Gupta, B.N. 1990a. *Statistics (Theory & Practice)*, Sahitya Bhawan, Agra, pp. 312-313, 515-584.

Gupta, M.D. 1990. Rehabilitating Jhumias On Rubber Plantations: An Assessment of the Tripura Experiment, in D.N. Majumdar (ed.) *Shifting Cultivation in North- East India*, Omsons Publications, New Delhi, pp. 308-311.

Gupta, M.D. 1994. Continuity and Change in Tribal Societies: A Study of the Economic Aspects of Tribal Jhumia Society in Tripura, in M.S. Sangma (ed.) *Essays On North-East India,* Indus Publishing Company, New Delhi, pp. 145-164.

Gupta, M.D. 2007. Planning for the Hill Economy of Tripura, in A. Banerjee, B. Kar (ed.) *Economic Planning and Development of North-Eastern States*, Kanishka Publishers, New Delhi, pp. 78-79.

Hasan, M.I. 2005. *Population Geography*, Rawat Publications, Jaipur, New Delhi, Bangalore, Hyderabad, Guwahati, Kolkata, pp. 40-73.

http://dhalai.gov.in/new/.html. Accessed on 3 June 2015. Israel, G. D. 1992. *Determining Sample Size*, PEOD-6, pp. 1-5.

Joseph, J., George K, T. and Dey, S.K. 2010. Report on the socio-economic impact of Natural Rubber cultivation under the block planting scheme in Tripura, *MPRA Paper No. 52370*, pp. 1-39.

Kamaruddin, R. and Samsudin, S. 2014. The Sustainable Livelihoods Index: A Tool to Assess the Ability and Preparedness of the Rural Poor in Receiving Entrepreneurial Project, *Journal of Social Economics Research, 1(6)*, pp. 108-117.

Kannan, M. 2013. Growth and Performance of Area, Production and Productivity of Natural Rubber in India, *International Journal of Research in Commerce, Economic & Management, Volume no.3, Issue no.10*, pp. 80-84.

Kesari, G.K. 2011. Geology and Mineral Resources of Manipur, Mizoram, Nagaland and Tripura, *Geological Survey of India, No. 30 Part IV, Vol. 1 (Part-2)*, Miscelleaneous Publication, pp. 8-10 and 78-82.

Khatun, D. and Roy, B.C. 2012. Rural Livelihood Diversification in West Bengal: Determinants and Constraints, *Agricultural Economics Research, Vol. 25 (No.1)*, pp. 115-124.

Kochhar, S.L. 1981. *Economic Botany in the Tropics*, Macmillan India Limited, Delhi, Bombay, Calcutta, Madras, pp. 385-395.

Krantz, L. 2001. *The Sustainable Livelihood Approach to Poverty Reduction- An Introduction*, Swedish International Development Cooperation Agency (Sida), Stockholm, Sweden, pp. 1-27.

Krishnakumar, A.K. and Meenattoor.J, R. 2003. Restoration of denuded lands in the north eastern India through rubber plantations, *LEISA INDIA*, December, pp. 23-24.

Kumar, A.K.K. 2014. Rubber Development Version 2.0 for North Eastern Region with a reference to Tripura, in K. Shib (ed.) *Rubber in Tripura (1963-2013)-Golden Jubilee Celebration, Souvenir*, Bina Printers, Arundhatinagar, Agartala, pp. 13-18.

Kurien, P.K. 2001. Sub-Marginal Rubber Cultivators: A study of livelihood issues of beneficiaries of 'Rubber to the Poor' project of Malanad Development Society, Kanjirappally, *Discussion Paper No. 32*, Kerala Research Programme on Local Level Development, Centre for Development Studies, Thiruvananthapuram, pp. 5-69.

Lagerqvist, Y. 2013. Imagining the Borderlands: Contending Stories of a Resource Frontier in Muang Sing. *Singapore Journal of Tropical Geography*, 34, pp. 57-69.

Laskar, S., Dadhwal, K.S. and Prasad, R.N. 1983. Soils of Tripura and their Fertility Management, *Research Bulletin No. 23*, Division of Soil Science, ICAR Research Complex for N.E.H. Region, Shillong (Tripura Centre), pp. 1-30.

Liu, W., Hu, H., Ma, Y. and Li, H. 2006. Environmental and Socioeconomic Impacts of Increasing Rubber Plantations in Menglun Township, Southwest China, *Mountain Research and Development, Vol. 26, No. 3,* pp. 245-253.

Luangmay, D. and Kaneko, S. 2013. Expansion of Rubber Tree Plantation in Northern Laos: Economic and Environmental Consequences, *Journal of International Development and Cooperation, Vol. 19, No. 3,* pp. 1-13.

Maithani, B.P. 2005. *Shifting Cultivation in North-East India: Policy issues and options,* Mittal Publication, New Delhi, pp. 30-50.

Meti, S., Pradeep, B., Jacob, J., Meerabai, M. and Jessy, M.D. 2014. Spatio- Temporal Analysis of Rubber Area and its Association with Soil and Topography in Kanyakumari District, *Rubber Science, 27(2),* pp. 182-192.

Meti, S., Pradeep, B., Jacob, J., Shebin, S.M. and Jessy, M.D. 2016. Application of Remote Sensing and GIS for Estimating Area under Natural Rubber Cultivation in India, *Rubber Science, 29(1),* pp. 7-19.

Mohanakumar, S. 2013. Expansion of Natural Rubber Cultivation in Tripura: Impact on Landholding, Employment and Income, *NLI Research Studies Series No. 107,* pp. 1-46.

Mohanakumar, S. 2014. Political Economy of Natural Rubber Cultivation in Tripura,

IDSJ Working Paper 172, pp. 1-15.

Mors, W.B. and Rizzini, C.T. 1966. *Useful Plants of Brazil,* Holden-Day Inc., San Francisco, London, Amsterdam, pp. 1-7.

Nath, D. and Bezbaruah, M.P. 2010. Income Augmentation in Small Farm Agriculture from Adoption of Rubber Production: A study in Northeast India, *YSA,* pp. 109-132.

Nath, T.K., Inoue, M. and Zoysa, M. 2013. Small-Scale Rubber Planting for Enhancement of People's Livelihoods: A Comparative Study in Three South Asian Countries, *Society and Naturl Resources: An International Journal, 26:9,* pp. 1066-1081.

Ochse, J.J., Soule, J., Dijkman, M.J. and Wehlburg, C. 1961. *Tropical and Subtropical Agriculture, Volume II,* The Macmillan Company, Collier- Macmillan Ltd., London, pp. 923-996.

Opeke, L.K. 1982. *Tropical Tree Crops,* John Wiley and Sons Ltd., Chichester, New York, Brisbane, Toronto, Singapore, pp. 216-240.

Panagos, P., Jones, A., Bosco, C. and Senthil Kumar, P.S. 2011, European digital archive on soil maps (EuDASM): preserving important soil data for public free access, *International Journal of Digital Earth, 4 (5),* 434-443.

Pensuk, A. and Shrestha, R.P. 2008. Effect of Paddy Area Conversion to Rubber Plantation on Rural Liveklihoods: A Case Study of Phatthalung Watershed, Southern Thailand, in W. Ongsakul (ed.) *GMSARN International Journal, Vol.2, No.4,* Greater Mekong Subregion Academic and Research Network, Thailand, pp. 185-190.

Peet, R.K. 1974. The Measurement of Species Diversity, *Annual Review of Ecology and Systematics, 5:1,* pp. 285-307.

Peter, K.V. 2002. *Plantation Crops,* National Book Trust, New Delhi, pp. 247-277.

Polhamus, L.G. 1962. *Rubber: Botany, Production, and Utilization,* Leonard Hill [Book] Limited, London, Interscience Publishers, INC., New York, pp. 1-30, 347-392.

Priyadarshan, P.M. 2011. *Biology of Hevea Rubber,* CAB International Publisher, pp.1-16.

Raghavan, K.K. 2009. Rubber for tribal welfare in Tripura, in G.K. Bera, B.K. Medhi,

R.P. Athparia, and K. Jose (ed.) *Tribal Development in Tripura,* EBH Publisher, Guwahati, pp. 89-98.

Rahman, S.A. 2006. *The Beautiful India- Tripura,* Reference Press Publication, New Delhi, pp. 191-200.

Rajasenan, D. 2010. Livelihood and employment of workers in rubber and spices plantations, *NRPPD Discussion Paper- 6,* pp. 1-68.

Raj, U., Hebbar, R., Ravishankar, H.M., Jacob, J., Ray, D., Meti, S., Shebin, S.M., & Pradeep, B. 2012. Geospatial Technology for Acreage Estimation of Natural

Rubber and Identification of Potential Areas for its Cultivation in Tripura, *Project Report, Issue no. 1,* National Remote Sensing Centre, ISRO, Govt. of India and Rubber Research Institute of India, Rubber Board, Min. of Commerce & Industry, Govt. of India, pp. 1-45.

Ranganath, B.K., PRADEEP, N., Manjula, V.B., Gowda, B., Rajanna, M.D., Shettigar, D., Rao, P.P.N. 2004. Detection of diseased rubber plantations using satellite remote sensing, *Journal of the Indian Society of Remote Sensing, Vol.32, No.1,* pp. 49-58.

Ray, D., Behera, M.D. and Jacob, J. 2014. Indian Brahmaputra valley offers significant potential for cultivation of rubber trees under change climate, *Current Science, Vol. 107, No. 3,* pp. 461-469.

Ray, D., Beehera, M.D.and Jacob, J. 2016. Predicting the distribution of rubber trees (Hevea brasiliensis) through ecological niche modeling with climate, soil, topography and socioeconomic factors, *Ecol Res, 31,* pp. 75-91.

Roy, P. 2005. *Economic Geography– A Study of Resources,* New Central Book Agency (P) Ltd., Kolkata ,pp. 205-209, 492.

Rubber Board. 2016. *Role of Rubber Board in Growth of Rubber Plantation in NE,* The regional office, Agartala.

Rubber Board. *The Rubber Tree.* Retrieved from http://rubberboard.org.in/rubbercultivation.asp. Accessed on April 2015.

Ruthenberg, H. 1980. *Farming systems in the tropics,* Oxford science publications-third edition, Oxford University Press, New York, Printed in Great Britain by The Thetford Press Limited, Thetford, Norfolk, pp. 301-305.

Saha, B. and Bahal, R. 2010. Livelihood Diversification Pursued by Farmers in West Bengal, *Indian Res. J. Ext. Edu., 10(2),* pp. 1-9.

Saha, B. and Bahal, R. 2014. Livelihood Diversification Pattern among the Farmers of West Bengal, *Economic Affairs, 59(3)*, pp. 321-334.

Saigal, O. 1978. *Tripura- Its History and Culture*, Concept Publishing Company, Delhi, pp. 17-27.

Sarkar, A.N. 1994. *Integrated Horticulture Development in Eastern Himalayas*, M.D.Publication Pvt. Ltd., New Delhi, pp. 114-123.

Sarkar, M. 1986. Forestry Development in Tripura During the Five Year Plans, in

M.D. Gupta, A.K. Gangopadhyay, T. Bhattacharya, M. Chakraborti (ed.) *Forestry Development in North-East India*, Omsons Publications, Guwahati and New Delhi, pp. 192-203.

Sarkar, S. 2011. Rubber Plantation: A new hope for rural tribals in Tripura,

International Journal of Plant Sciences, Vol. 6 Issue 2, pp. 274-276.

Sharma, G., Joseph, J., George K, T. and Dey, S.K. 2011. Impact of Mahatma Gandhi National Rural Employment Guarantee Act on Rubber Block Plantation Scheme in Tripura, *Agriculture Economic Research Review Vol. 24*, pp. 525- 530.

Sharma, G., George K., T. and Dey, S.K. 2014. Natural Rubber Sector in Tripura: Role of Institutional Innovations, *Agricultural Economics Research Review, Vol. 27*, pp 17-24.

Sinha, A.K. 2010. Prospect of rubber plantation in NE region with special reference to the state of Tripura, *The Tripura Foundation*. Retrieved from http://thetripurafoundation.org/art-rubber-prospect-tripura-and-north-east.

Accessed on April 2015.

Sinha, A. K. 2012. Rubber plantation in Northeast India, hopes vs concerns.

The Tripura Foundation. Retrieved from http://thetripurafoundation.org/art-rubber-north-east-hopes-vs-concern. Accessed on April 2015.

Simpson, E.H. 1949. Measurement of Diversity, *Nature, Vol. 163*, p. 688.

Sun, Z., Leinenkugel, P., Guo, H., Huang, C. and Kuenzer, C. 2017. Extracting distribution and expansion of rubber plantations from Landsat imagery using the C5.0 decision tree method, *Journal of Applied Remote Sensing, 11(2)*, pp. 1-21.

TFDPCL, 2016. *Documents on TFDPCL*, TFDPC Bhavan, Agartala, Tripura. TFDPCL. 2016a. *Plan for Responsible Rubber wood and Bamboo Plantations*

Management (2013-14 to 2017-18), Agartala, Tripura, Prepared in consultation with Indian Institute of Forest Management, NEHRU NAGAR, BHOPAL, pp. 1-201.

Tilahun and Teferie. 2015. Accuracy Assessment of Land Use Land Cover Classification using Google Earth, *American Journal of Environment Protection, Vol. 4, No. 4*, pp. 193-198.

Tiwari, R.C. 2008. *Geography of India*, Prayag Pustak Bhawan, Allahabad, pp. 344-346.

Tripura, B. 2013. Tribal Question in Tripura: Dialogue Between Its Past and Present,

Journal of Tribal Intellectual Collective India, Vol. 1 Issue 2 No 3, pp. 33-50. TRPCL. 2012. *Brief Note on Tripura Rehabilitation Plantation Corporation Ltd.*

Diary 2012, Agartala, Tripura, pp. 1-12.

TRPCL. 2016. *Document on Tripura Rehabilitation Plantation Corporation Limited*, Agartala, Tripura.

USGS. EarthExplorer. Available online: http://earthexplorer.usgs.gov/.

Verheye, W. 2010. Growth and Production of Rubber, in Verheye, W. (ed.) *Land Use, Land Cover and Soil Sciences*, Encyclopedia of Life Support Systems (EOLSS), UNESCO-EOLSS Publishers, Oxford, UK, pp. 1-20.

Vinod, K.K., Meenottoor, J.P., Priyadarshan, P.M., Pothen, J., Chaudhuri, D., Krishnakumar, A.K., Sethuraj, M.R., Potty, S.N. 1996. Early performance of some clones of Hevea Brasiliensis in Tripura, *Indian Journal of Natural Rubber Research, 9(2)*, pp. 123-129.

Vinod, K.K., Meenattoor, J.R., Reddy, Y.A., Priyadarshan, P.M., Chaudhuri, D. 2010. Ontogenetic variations in flush development are indicative of low temperature tolerance in *Hevea brasiliensis* clones, *Annals of Forest Research, 53(2)*, pp. 95-105.

Viswanathan, P.K. 2008. Emerging Smallholder Rubber Farming System in India and Thailand: A Comparative Economic Analysis, *Asian Journal of Agriculture and Development, Vol. 5, No. 2*, pp. 1-16.

Viswanathan, P.K. and Bhowmik, I. 2014. Compatibility of Institutional Architecture for Rubber Plantation Development in North East India from the Comparative Perspective of Kerala, *NRPPD Discussion Paper No. 38*, Centre For Development Studies, Thiruvananthapuram, pp. 1-71.

Viswanathan, P.K. and Shivakoti, G.P. 2008. Adoption of rubber-integrated farm livelihood systems: contrasting empirical evidence from the Indian context. *J For Res, 13*, pp. 1-14.

Vongkhamheng, C., Zhou, J.H., Beckline, M. and Phimmachanh, S. 2016. Socioeconomic and Ecological Impact Analysis of Rubber Cultivation in Southeast Asia. *Open Access Library Journal*, 3: e2339.

Vongkhamor, S., Phimmasen, K., Silapeth, B., Xayxomphou, B. and Petterson, E. 2007. *Key Issues in Smallholder Rubber Planting in Oudomxay and Luang Prabang Provinces*, National Agriculture and Forestry Research Institute Vientiane, Lao PDR.

World Resources Institute (WRI) in collaboration with United Narions Development Programme, United Nations Environment Programme, and World Bank., 2005. World Resources 2005: *The Wealth of the Poor-Managing Ecosystems to Fight Poverty*, World Resources Institute, Washington, DC, pp. 3- 246.

Xiaona, L., Zhiming, F., Luguang, J., Peng, L., Chenhua, L., Yanzhao, Y. and Zhen,Y. 2013. Rubber plantation and its relationship with topographical factors in the border region of China, Laos and Myanmar, *Journal of Geographical Sciences, Vol 23 (6)*, pp. 1019-1040.

Appendices

Appendix - 1

Annual gain of area under rubber plantation in Tripura (1972-73 to 2015-16)

Year	Total area under rubber plantation (in ha)	Annual gain in area under rubber plantation (in ha)	Annual growth rate of area under rubber plantation (in %)	Trend in the increase in annual gain of area under rubber plantation (in ha) (3 year moving average method)
1972-73	103			
1973-74	223	120	116.50	
1974-75	307	84	37.67	105
1975-76	417	110	35.83	117
1976-77	574	157	37.65	190
1977-78	878	304	52.96	296
1978-79	1,306	428	48.75	472
1979-80	1,989	683	52.30	623
1980-81	2,746	757	38.06	761
1981-82	3,590	844	30.74	863
1982-83	4,578	988	27.52	864
1983-84	5,338	760	16.60	985
1984-85	6,545	1,207	22.61	1,055
1985-86	7,742	1,197	18.29	1,582
1986-87	10,085	2,343	30.26	2,042
1987-88	12,670	2,585	25.63	2,195
1988-89	14,326	1,656	13.07	2,178
1989-90	16,620	2,294	16.01	1,550
1990-91	17,320	700	4.21	1,178
1991-92	17,860	540	3.12	543
1992-93	18,250	390	2.18	450
1993-94	18,670	420	2.30	464
1994-95	19,252	582	3.12	837
1995-96	20,761	1,509	7.84	1,104

Year	Total area under rubber plantation (in ha)	Annual gain in area under rubber plantation (in ha)	Annual growth rate of area under rubber plantation (in %)	Trend in the increase in annual gain of area under rubber plantation (in ha) (3 year moving average method)
1996-97	21,982	1,221	5.88	1,110
1997-98	22,582	600	2.73	1,120
1998-99	24,120	1,538	6.81	1,133
1999-00	25,380	1,260	5.22	1,304
2000-01	26,495	1,115	4.39	1,276
2001-02	27,947	1,452	5.48	1,158
2002-03	28,853	906	3.24	1,425
2003-04	30,770	1,917	6.64	1,373
2004-05	32,065	1,295	4.21	1,779
2005-06	34,189	2,124	6.62	1,663
2006-07	35,760	1,571	4.60	2,535
2007-08	39,670	3,910	10.93	4,133
2008-09	46,588	6,918	17.44	5,177
2009-10	51,292	4,704	10.10	4,666
2010-11	53,669	2,377	4.63	3,678
2011-12	57,621	3,952	7.36	3,313
2012-13	61,231	3,610	6.27	2,953
2013-14	62,529	1,298	2.12	4,225
2014-15	70,296	7,767	12.42	4,368
2015-16	74,335	4,039	5.75	

Source: Centre for Development Studies; Rubber Board, 2016 (gain in area under rubber plantation, growth rate, and 3 year moving average computed by the researcher).

Appendix - 2

District wise distribution of area under rubber plantation in Tripura (2008-09 to 2015-16)

Year	Total area under rubber plantation (in ha) in North Tripura district	Total area under rubber plantation (in ha) in Dahlia district	Total area under rubber plantation (in ha) in West Tripura district	Total area under rubber plantation (in ha) in South Tripura district
2008-09	5,324	2,978	21,486	16,800
2009-10	5,998	3,380	24,024	17,890
2010-11	6,256	3,451	25,088	18,874
2011-12	6,917	4,074	27,020	19,610

Year	Total area under rubber plantation (in ha) in North Tripura district	Total area under rubber plantation (in ha) in Dahlia district	Total area under rubber plantation (in ha) in West Tripura district	Total area under rubber plantation (in ha) in South Tripura district
2012-13	7,442	4,740	29,069	19,980
2013-14	7,445	5,151	29,869	20,064
2014-15	9,365	3,874	35,317	21,740
2015-16	9,759	4,230	37,479	22,867

Source: Rubber Board, cited by DESPD (2008-09, 2009-10, 2010-11, 2011-12, 2012-13, 2013-14, 2014-15, 2015-16); Rubber Board, 2016.

Appendix - 3

District wise growth rate of area under rubber plantation in Tripura (2008-09 to 2015-16)

Year	Total area under rubber plantation (in ha) in North Tripura district	Total area under rubber plantation (in ha) in Dhalai district	Total area under rubber plantation (in ha) in West Tripura district	Total area under rubber plantation (in ha) in South Tripura district
2008-09	5,324	2,978	21,486	16,800
2009-10	5,998 (12.66)	3,380 (13.50)	24,024 (11.81)	17,890 (6.49)
2010-11	6,256 (4.30)	3,451 (2.10)	25,088 (4.43)	18,874 (5.50)
2011-12	6,917 (10.57)	4,074 (18.05)	27,020 (7.70)	19,610 (3.90)
2012-13	7,442 (7.59)	4,740 (16.35)	29,069 (7.58)	19,980 (1.89)
2013-14	7,445 (0.04)	5,151 (8.67)	29,869 (2.75)	20,064 (0.42)
2014-15	9,365 (25.79)	3,874 (-24.79)	35,317 (18.24)	21,740 (8.35)
2015-16	9,759 (4.21)	4,230 (9.19)	37,479 (6.12)	22,867 (5.18)

Source: Rubber Board, cited by DESPD (2008-09, 2009-10, 2010-11, 2011-12, 2012-13, 2013-14, 2014-15, 2015-16); Rubber Board, 2016 (figure in parenthesis indicate annual growth rate (in %) in area under rubber plantation computed by the researcher).

Appendix - 4

Share of area under rubber plantation developed by the TFDPCL to the total area under rubber plantation of Tripura (1976-77 to 2015-16)

Year	Total area under rubber plantation (in ha) in the state	Total area under rubber plantation (in ha) of the TFDPCL	Share (in %) of area under rubber plantation of the TFDPCL to the total area under rubber plantation of the state
1976-77	574	148	25.78
1977-78	878	451	51.37
1978-79	1,306	867	66.39

Year	Total area under rubber plantation (in ha) in the state	Total area under rubber plantation (in ha) of the TFDPCL	Share (in %) of area under rubber plantation of the TFDPCL to the total area under rubber plantation of the state
1979-80	1,989	1,534	77.12
1980-81	2,746	2,232	81.28
1981-82	3,590	2,948	82.12
1982-83	4,578	3,673	80.23
1983-84	5,338	4,190	78.49
1984-85	6,545	4,825	73.72
1985-86	7,742	5,361	69.25
1986-87	10,085	6,265	62.12
1987-88	12,670	7,169	56.58
1988-89	14,326	7,646	53.37
1989-90	16,620	8,620	51.87
1990-91	17,320	8,709	50.28
1991-92	17,860	8,840	49.50
1992-93	18,250	8,971	49.16
1993-94	18,670	9,059	48.52
1994-95	19,252	9,158	47.57
1995-96	20,761	9,390	45.23
1996-97	21,982	9,479	43.12
1997-98	22,582	9,537	42.23
1998-99	24,120	9,566	39.66
1999-00	25,380	9,614	37.88
2000-01	26,495	9,757	36.83
2001-02	27,947	9,787	35.02
2002-03	28,853	9,929	34.41
2003-04	30,770	10,093	32.80
2004-05	32,065	10,192	31.79
2005-06	34,189	10,214	29.88
2006-07	35,760	10,594	29.63
2007-08	39,670	10,877	27.42
2008-09	46,588	11,040	23.70
2009-10	51,292	11,083	21.61
2010-11	53,669	11,134	20.75

Year	Total area under rubber plantation (in ha) in the state	Total area under rubber plantation (in ha) of the TFDPCL	Share (in %) of area under rubber plantation of the TFDPCL to the total area under rubber plantation of the state
2011-12	57,621	11,181	19.40
2012-13	61,231	11,201	18.29
2013-14	62,529	11,203	17.92
2014-15	70,296	11,211	15.95
2015-16	74,335	11,624	15.64

Source: Centre for Development Studies; Rubber Board, 2016; TFDPCL, 2016 (Share of area under rubber plantation computed by the researcher).

Appendix - 5

Annual gain of area under rubber plantation developed by the TFDPCL (1976-77 to 2015-16)

Year	Total area under rubber plantation (in ha)	Annual gain of area under rubber plantation (in ha)	Annual growth rate of area under rubber plantation (in %)	Trend in the increase in annual gain of area under rubber plantation (in ha) (3 year moving average method)
1976-77	148	148	0.00	
1977-78	451	303	204.73	289
1978-79	867	416	92.24	462
1979-80	1,534	667	76.93	594
1980-81	2,232	698	45.50	694
1981-82	2,948	716	32.08	713
1982-83	3,673	725	24.59	653
1983-84	4,190	517	14.08	626
1984-85	4,825	635	15.16	563
1985-86	5,361	536	11.11	692
1986-87	6,265	904	16.86	781
1987-88	7,169	904	14.43	762
1988-89	7,646	477	6.65	785
1989-90	8,620	974	12.74	513
1990-91	8,709	89	1.03	398
1991-92	8,840	131	1.50	117
1992-93	8,971	131	1.48	117
1993-94	9,059	88	0.98	106
1994-95	9,158	99	1.09	140

Year	Total area under rubber plantation (in ha)	Annual gain of area under rubber plantation (in ha)	Annual growth rate of area under rubber plantation (in %)	Trend in the increase in annual gain of area under rubber plantation (in ha) (3 year moving average method)
1995-96	9,390	232	2.53	140
1996-97	9,479	89	0.95	126
1997-98	9,537	58	0.61	59
1998-99	9,566	29	0.30	45
1999-00	9,614	48	0.50	73
2000-01	9,757	143	1.49	74
2001-02	9,787	30	0.31	105
2002-03	9,929	142	1.45	112
2003-04	10,093	164	1.65	135
2004-05	10,192	99	0.98	95
2005-06	10,214	22	0.22	167
2006-07	10,594	380	3.72	228
2007-08	10,877	283	2.67	275
2008-09	11,040	163	1.50	163
2009-10	11,083	43	0.39	86
2010-11	11,134	51	0.46	47
2011-12	11,181	47	0.42	39
2012-13	11,201	20	0.18	23
2013-14	11,203	2	0.02	10
2014-15	11,211	8	0.07	141
2015-16	11,624	413	3.68	

Source: TFDPCL, 2016 (gain in area under rubber plantation, growth rate, and 3 year moving average computed by the researcher).

Appendix - 6

Share of area under rubber plantation developed by the TRPCL to the total area under rubber plantation of Tripura (1984-85 to 2015-16)

Year	Total area under rubber plantation (in ha) in the state	Total area under rubber plantation (in ha) of the TRPCL	Share (in %) of area under rubber plantation of the TRPCL to the total area under rubber plantation of the state
1984-85	6,545	75	1.15
1985-86	7,742	245	3.16
1986-87	10,085	580	5.75

Year	Total area under rubber plantation (in ha) in the state	Total area under rubber plantation (in ha) of the TRPCL	Share (in %) of area under rubber plantation of the TRPCL to the total area under rubber plantation of the state
1987-88	12,670	1,072	8.46
1988-89	14,326	1,449	10.11
1989-90	16,620	2,236	13.45
1990-91	17,320	2,352	13.58
1991-92	17,860	2,476	13.86
1992-93	18,250	2,608	14.29
1993-94	18,670	2,711	14.52
1994-95	19,252	2,851	14.81
1995-96	20,761	3,000	14.45
1996-97	21,982	3,084	14.03
1997-98	22,582	3,299	14.61
1998-99	24,120	3,358	13.92
1999-00	25,380	3,585	14.13
2000-01	26,495	3,870	14.61
2001-02	27,947	4,237	15.16
2002-03	28,853	4,505	15.61
2003-04	30,770	4,905	15.94
2004-05	32,065	5,095	15.89
2005-06	34,189	5,496	16.08
2006-07	35,760	6,079	17.00
2007-08	39,670	6,668	16.81
2008-09	46,588	7,278	15.62
2009-10	51,292	7,689	14.99
2010-11	53,669	7,816	14.56
2011-12	57,621	7,952	13.80
2012-13	61,231	8,186	13.37
2013-14	62,529	8,372	13.39
2014-15	70,296	8,780	12.49
2015-16	74,335	9,154	12.31

Source: Centre for Development Studies; Rubber Board, 2016; TRPCL, 2016 (Share of area under rubber plantation computed by the researcher).

Appendix – 7

Annual gain of area under rubber plantation developed by the TRPCL (1984-85 to 2015-16)

Year	Total area under rubber plantation (in ha)	Annual gain of area under rubber plantation (in ha)	Annual growth rate of area under rubber plantation (in %)	Trend in the increase in annual gain of area under rubber plantation (in ha) (3 year moving average method)
1984-85	75	75	0.00	
1985-86	245	170	226.67	193
1986-87	580	335	136.73	332
1987-88	1,072	492	84.83	401
1988-89	1,449	377	35.17	552
1989-90	2,236	787	54.31	427
1990-91	2,352	116	5.19	342
1991-92	2,476	124	5.27	124
1992-93	2,608	132	5.33	120
1993-94	2,711	103	3.95	125
1994-95	2,851	140	5.16	131
1995-96	3,000	149	5.23	124
1996-97	3,084	84	2.8	149
1997-98	3,299	215	6.97	119
1998-99	3,358	59	1.79	167
1999-00	3,585	227	6.76	190
2000-01	3,870	285	7.95	293
2001-02	4,237	367	9.48	307
2002-03	4,505	268	6.32	345
2003-04	4,905	400	8.88	286
2004-05	5,095	190	3.87	330
2005-06	5,496	401	7.87	391
2006-07	6,079	583	10.61	524
2007-08	6,668	589	9.69	594
2008-09	7,278	610	9.15	537
2009-10	7,689	411	5.65	383
2010-11	7,816	127	1.65	225
2011-12	7,952	136	1.74	166
2012-13	8,186	234	2.94	185
2013-14	8,372	186	2.27	276
2014-15	8,780	408	4.87	323
2015-16	9,154	374	4.26	

Source: TRPCL, 2016 (3 year moving average computed by the researcher).

Appendix – 8

Share of area under rubber plantation developed by the Rubber Board to the total area under rubber plantation of Tripura (1992-93 to 2015-16)

Year	Total area under rubber plantation (in ha) in the state	Total area under rubber plantation (in ha) developed by the Rubber Board	Share (in %) of area under rubber plantation developed by the Rubber Board to the total area under rubber plantation of the state
1992-93	18,250	122	0.67
1993-94	18,670	337	1.81
1994-95	19,252	623	3.24
1995-96	20,761	1,204	5.80
1996-97	21,982	1,356	6.17
1997-98	22,582	1,602	7.09
1998-99	24,120	2,049	8.50
1999-00	25,380	2,220	8.75
2000-01	26,495	2,346	8.85
2001-02	27,947	2,710	9.70
2002-03	28,853	2,806	9.73
2003-04	30,770	3,073	9.99
2004-05	32,065	3,176	9.90
2005-06	34,189	3,301	9.66
2006-07	35,760	3,402	9.51
2007-08	39,670	3,402	8.58
2008-09	46,588	3,564	7.65
2009-10	51,292	3,620	7.06
2010-11	53,669	3,725	6.94
2011-12	57,621	3,725	6.46
2012-13	61,231	3,775	6.17
2013-14	62,529	3,775	6.04
2014-15	70,296	3,775	5.37
2015-16	74,335	3,775	5.08

Source: Centre for Development Studies; Rubber Board, 2016 (Share of area under rubber plantation computed by the researcher).

Appendix – 9

Annual gain of area under rubber plantation developed by the Rubber Board (1984-85 to 2015-16)

Year	Total area under rubber plantation (in ha)	Annual gain of area under rubber plantation (in ha)	Annual growth rate of area under rubber plantation (in %)	Trend in the increase in annual gain of area under rubber plantation (in ha) (3 year moving average method)
1992-93	122	122	0.00	
1993-94	337	215	176.23	208
1994-95	623	286	84.87	361
1995-96	1,204	581	93.26	340
1996-97	1,356	152	12.62	326
1997-98	1,602	246	18.14	282
1998-99	2,049	447	27.90	288
1999-00	2,220	171	8.35	248
2000-01	2,346	126	5.68	220
2001-02	2,710	364	15.52	195
2002-03	2,806	96	3.54	242
2003-04	3,073	267	9.52	155
2004-05	3,176	103	3.35	165
2005-06	3,301	125	3.94	110
2006-07	3,402	101	3.06	75
2007-08	3,402	0	0.00	88
2008-09	3,564	162	4.76	73
2009-10	3,620	56	1.57	108
2010-11	3,725	105	2.90	54
2011-12	3,725	0	0.00	52
2012-13	3,775	50	1.34	17
2013-14	3,775	0	0.00	17
2014-15	3,775	0	0.00	0
2015-16	3,775	0	0.00	

Source: Rubber Board, 2016 (3 year moving average computed by the researcher).

Appendix - 10

RPC/village wise indicator suggesting human asset/capital of the BHHs

RPC/ village Indicator	Depacherra		Dakshin Dhumacherra		Pathaliaghat		Dakshin Taidu	
	Share (in %) of the BHHs (n=54)	0-1 score value	Share (in %) of the BHHs (n=45)	0-1 score value	Share (in %) of the BHHs (n=56)	0-1 score value	Share (in %) of the BHHs (n=55)	0-1 score value
Working population	70.68	0.07	70.19	0.00	67.21	0.13	64.15	0.00
Literacy rate	70.87	0.08	81.77	0.39	83.90	0.57	74.17	0.31
Traditional skills known by the members of the household	68.52	0.00	73.33	0.11	62.50	0.00	69.09	0.15
Modern skills known by the members of the household	98.15	1	100	1	100	1	96.36	1
Composite index score of human asset/ capital		0.29		0.38		0.43		*0.37*

Source: Field survey, 2016.

Appendix – 11

RPC/village wise indicator suggesting social asset/capital of the BHHs

RPC/ village Indicator	Depacherra		Dakshin Dhumacherra		Pathaliaghat		Dakshin Taidu	
	Share (in %) of the BHHs (n=54)	0-1 score value	Share (in %) of the BHHs (n=45)	0-1 score value	Share (in %) of the BHHs (n=56)	0-1 score value	Share (in %) of the BHHs(n=55)	0-1 score value
Good relationship with the TRPCL official	100	1	100	1	100	1	100	1
Training and capacity development activities	64.81	0.60	60.00	0.60	57.14	0.52	54.55	0.55
Committee member in the locality	12.96	0.02	13.33	0.13	14.29	0.04	9.09	0.09
Member in Self Help Group (SHG)	11.11	0.00	0.00	0.00	10.71	0.00	0.00	0.00
Participated in programme or meeting	74.07	0.71	97.78	0.98	94.64	0.94	90.91	0.91
Composite index score of social asset/ capital		0.47		0.54		0.50		0.51

Source: Field survey, 2016.

Appendix – 12

RPC/village wise indicator suggesting natural asset/capital of the BHHs

RPC/ village Indicator	Depacherra		Dakshin Dhumacherra		Pathaliaghat		Dakshin Taidu	
	Share (in %) of the BHHs (n=54)	0-1 score value	Share (in %) of the BHHs (n=45)	0-1 score value	Share (in %) of the BHHs (n=56)	0-1 score value	Share (in %) of the BHHs (n=55)	0-1 score value
Landholding size below 1 ha	20.37	0.17	2.22	0.00	37.50	0.66	10.91	0.12
Landholding size between 1 – 1.5 ha	75.93	0.75	80.00	1	57.14	1	61.82	1
Landholding size above 1.5 ha	3.70	0.00	17.78	0.20	5.36	0.09	27.27	0.41
Access to land for *jhum* cultivation	5.56	0.02	8.89	0.09	0.00	0.00	3.64	0.00
Access to water sources (Pond/ Stream/River)	35.19	0.33	60.00	0.74	10.71	0.19	20.00	0.28
Access to common forest resources	100	1	57.78	0.71	14.29	0.25	50.91	0.81
Composite index score of natural asset/capital		0.38		0.46		0.37		0.44

Source: Field survey, 2016.

Appendix – 13

RPC/village wise indicator suggesting physical asset/capital of the BHHs

RPC/ village Indicators	Depacherra		Dakshin Dhumacherra		Pathaliaghat		Dakshin Taidu	
	Share (in %) of the BHHs (n=54)	0-1 score value	Share (in %) of the BHHs (n=45)	0-1 score value	Share (in %) of the BHHs (n=56)	0-1 score value	Share (in %) of the BHHs (n=55)	0-1 score value
Access to rubber processing centre	100	1	100	1	100	1	100	1
Access to blacktop road within a kilometre (km)	100	1	100	1	100	1	100	1
Access to market within 5 km	81.48	0.00	0.00	0.00	100	1	100	1
Access to safe drinking water source and sanitation	100	1	100	1	100	1	100	1
Access to school	100	1	100	1	100	1	100	1
Access to electricity	100	1	100	1	100	1	100	1
Composite index score of physical asset/capital		0.83		0.83		1		1

Source: Field survey, 2016.

Appendix – 14

RPC/village wise indicator suggesting financial asset/capital of the BHHs

RPC/ village Indicator	Depacherra		Dakshin Dhumacherra		Pathaliaghat		Dakshin Taidu	
	Share (in %) of the BHHs (n=54)	0-1 score value	Share (in %) of the BHHs (n=45)	0-1 score value	Share (in %) of the BHHs (n=56)	0-1 score value	Share (in %) of the BHHs (n=55)	0-1 score value
Rs 3,000-6,000	0.00	0.00	4.44	0.05	0.00	0.00	5.45	0.09
Rs 6,001-9,000	64.81	1	48.89	1	53.57	1	41.82	1
Rs 9,001-12,000	27.78	0.43	28.89	0.57	30.36	0.57	32.73	0.77
Rs 12,001-15,000	7.41	0.11	15.56	0.29	14.29	0.27	18.18	0.41
Rs 15,001-18,000	0.00	0.00	2.22	0.00	1.79	0.03	1.82	0.00
Composite index score of financial asset/capital		0.31		0.38		0.37		0.45

Source: Field survey, 2016.

Appendix – 15

HOUSEHOLD SCHEDULE

Questions for households rehabilitated through rubber plantation

This questionnaire intends to collect some socio-economic information of those households who are the beneficiary of the rubber plantation based rehabilitation programme. The information provided by the household for this survey will in no case deprive any household. The information collected as part of this questionnaire will be used for academic purpose. Honest cooperation in this process is solicited for this purpose.

Date............./............../2016

Identification and Basic information:

1. Name of the Head of the Household/Respondent: ....
2. Name of the District and Sub-division: ....
3. Name of the Block: ....
4. Name of the RPC/village: ....
5. Total number of households in the village: ....
6. Number of households rehabilitated though rubber plantation scheme in the village: ....

Section- A

A. Human capital

1. Total members of the Household: ....

Male (M) Female (F)	Age	Education Qualification	Occupation /major livelihood	Any other	Gender:
					1. Male (M) 2. Female (F) **Education:** 1. illiterate 2. nursery 3. primary (I to iv) 4. elementary (v to vii) 5. secondary (viii to x) 6. higher secondary (xi to xii) 7. graduate 8. post graduate 9. other mention it **Major livelihoods** 1 Tapper 2. Latex collector 3. Processer of latex 4. Both 1 & 2 5. Wage earner 6-employed (pvt/ govt) 7- other mention it **Business:** 1. Shop 2. Vendor 3. Driver 4. other mention it

2. **Religion:** 1.Hinduism, 2.Christianity, 3.Islam, 4.Others ....
3. **Community/Tribe:** 1.Tripura, 2.Reang, 3.Debbarma, 4.Jamatia, 5.any other ....
4. **House type:** 1.Kaccha (mud wall with tin roof), 2.Half wall with bamboo wall with tin roof, 3. Brick (pucca) wall with tin roof, 4.RCC (pucca), 5.Others ....

5. **Type of Sanitation:** 1. Pucca, 2. Mud wall, 3. Bamboo, 4. Open, 5. Piped sewer system, 6. Septic tank, 7. Pit, 8-Public, 9. Others ....
6. **What modern technologies and training connected with rubber are you having access to:** ....
7. **What other modern technique or skills do you or any members of the households know/practices:** ....
8. **Do you** or **any members of the household still know/ practice traditional Skills?:** 1. Yes..........................2. No..............................if **Yes** like: 1. Handicraft, 2. Weaving, 3. Pottery, 4. Folk artist, 5. Priest, 6- Shifting cultivation, 7. Fishing & hunting and 6- others...
9. Rearing of livestock?: 1. Yes............... 2. No...............if Yes like: 1. Rearing hen, 2. Rearing pig, 3. Rearing cattle, 4. Rearing goat, 5. Rearing duck, 6. other................................
10. **What are the main sources of livelihood/income of you or the household in the past:** ....

B. Financial capital

1. **Total Monthly/Annual income of the household:** Rs...
2. **Monthly/Annual income from rubber plantation:** Rs..
3. **Monthly /Annual income from (if) working in other rubber plantation:** Rs.. ....
4. **Monthly/Annual income, if employed in rubber processing centre:** Rs......... ....
5. **Monthly/Annual income from non-farm activities like:** 1. **Shop** Rs.................. 2. **Driver** Rs...................3. **Decorator** Rs...................4. **Mechanic** Rs.................., 5. **Carpentry** Rs................,6. **Petty business in local market** Rs..........................

B.1. Expenditure

1. **Monthly/Annual expenditure in education of children: Rs** ....
2. **Monthly/Annual household expenditures on food items:** Rs...........................
3. **Annual Expenditure on medical treatment:** Rs..
4. **Annual Expenditure on others household articles: Rs**.......................................

C. Natural capital

1. **Total land holding size of the household in kani/hectare:** ....
2. **Size of rubber cultivated land in kani/hectare:** ....
3. **Nature of land ownership:** 1.Owner, 2.Moratage to the Corporation, and 3.any other..
4. **Beside rubber plantation, land under:** 1. Kitchen garden, 3. Wet paddy, 4. Pond, 5. any other...

5. **Access to land for shifting cultivation: 1.** Yes................ **2.** No.................
6. **Access to forest area for collection of household items/needs: 1.**Yes............. **2.** No.................
7. **Access to stream and water bodies: 1.** Yes............... **2.** No..............
8. **Do you/household have fish pond for growing fishery: 1.** Yes........ **2.** No............
9. **Has the value of land increased/decreased with the introduction of rubber plantation: 1**. Yes............ **2.** No...............

D. Physical capital

1. **Access to road:** 1.blacktop road, 2.brick road, and 3.mud/kucha road
2. **Access to infrastructures facilities like:** 1.rubber processing & storage centre, 2.training and capacity building, 3.rubber board office, 4.All of these ....
3. **Access to any Research and development infrastructure facilities:** 1. Yes............... 2. No..................
4. **Access to market:** 1. within 5 Km, 2.More than 5 Km
5. **Access to school:** 1. Yes................ **2.** No..............
6. **Access of electricity in the house:** 1. Yes............... **2.** No...............
7. **Access to safe drinking water source:** 1. Hand pump, 2. Water supply, 3. pumped well, 4.well, 5.any others ....
8. **Safe drinking water asset within the House:** 1. Yes.............. 2. No............if **Yes** like i. water filter, ii-any other..
9. **Animals assets owned:** 1. Hens, 2.goats, 3.pigs, 4.ducks, 5.cattle, 6-any other .. ..
10. **Mention other assets owned by you/the household:** ....

E. Social capital

1. **Relationship with the neighbour families:** ....
2. **Relationship with the officials of the TRPCL:** ....
3. **Access to (are you part of):** 1. Rubber Grower Societies (RGS), 2. Rubber Producer Societies (RPS), 3. Beneficiary committee member, 4. rubber marketing societies, 5. Self help groups, 6. Any other..
4. **Do you receive any help and facilities from the corporation/agency:** 1. Yes................ 2. No..................
5. **Access to any local development institutions provided by the govt.** 1. Yes............. 2. No ....
6. **Access to training and capacity development activities:** 1. Yes........... 2. No........... if Yes like 1. Research & Development facilities, 2. training in tapping, 3. training in processing, 4. Any other...

7. **Part of committee members in the locality:** 1. Yes....2.No.....
8. **Are there any (access to) groups/organizations/networks in the local area other the rubber growing agencies:** 1. Yes.................. 2. No................... if **Yes** mention......................
9. **Participation in programme or meeting organized by rubber growing agencies and others government and non-government organization regarding rubber and others:** 1. Yes................ 2. No.................
10. **Apart from rubber plantation development scheme, do any other government schemes or policies (like MGNREGA, Indira Awas Yogana, National rural livelihood mission, Prime Minister's Employment Generation Programme, SWABALAMAN Programme of Government of Tripura etc) implemented in the area to (strengthening or undermining) the livelihoods of the local communities:** 1. Yes................. 2. No................ If **Yes** mention ....

Section- B

Group discussion

1. **In how many *kani* or hectare rubber plantation is usually cultivated per family: ....**
2. **Was other crops grown along with the rubber trees:** 1. Yes................ 2. No .... if Yes like 1. Banana, 2. Pineapple, 3. Ginger and Turmeric, 4. Vegetables, 5. Medicinal plants, 6. Tuber crops and 7. Any other..
3. **Is beekeeping practice in the rubber plantation:** 1. Yes.............. 2. No.............
4. **Is there any conflicts/disputes between the rubber growing agencies and the local people regarding rubber cultivation in the area:** 1. Yes............... 2. .... No. if Yes mention the reason ....
5. **Is rain guarding protective measure use for latex during rainy and winter season: 1.**Yes................... **2.** No ....
6. **During latex collection do any protective measure is put on: 1.** Yes............. **2.** No....................

Livelihood options before rubber plantations were introduced-

****What are the main sources of livelihood/income of you or the household before rubber plantation was introduce: ....**

1. **Do you own any land before introduction of rubber:** 1. Yes......... 2. No...........
2. **Monthly/Annual income of the household before rubber was introduced:** Rs..
3. **Main source of earning before rubber was introduced like:** 1. wage labour, 2. *jhum* cultivation, 3. Any other ....
4. **Attitude and habits like:** 1. Vulnerability/risk of being harm, 2. Misguides by anti-socials, 3. Any other ....
5. **How was the standard of education before introduction of rubber plantation?:** 1. Good .... 2. Poor .... 3. large scale illiteracy ....
6. **How was the well-being/welfare of the households: 1.** Good............... **2. ....** Poor................... **3.** Very poor/bad ....
7. **How was the infrastructure facilities of the areas: 1.** Good.................. **2.** Poor................. **3.** Very poor/bad ....

About rubber plantation development scheme

1. **What improvement has the rubber plantation development scheme brought to the household, the community and the area:**
2. **How is the area now after the implementation of rubber plantation?:**
3. **Any problem associated with rubber plantation development scheme:**

1. Yes................ **2.** No .... if **Yes** what changes and priorities and relevant initiatives can be brought in this rubber plantation development scheme in the area:

4. **What services and activities would you like the rubber corporation and the state government to provide for present and future betterment of the household, the community and the area:**
5. **What were your view and opinion about rubber plantation and its scheme in the beginning/before and what do you view and feel now about rubber plantation?:**

Appendix – 16

HOUSEHOLD SCHEDULE

Questions for households not having rubber plantation

This questionnaire intends to collect some socio-economic information of those households who are the beneficiary of the rubber plantation based rehabilitation programme. The information provided by the household for this survey will in no case deprive any household. The information collected as part of this questionnaire will be used for academic purpose. Honest cooperation in this process is solicited for this purpose.

Date: ..../......................../2016

Identification and Basic information:

1. Name of the Head of the Household/Respondent: ....
2. Name of the District and Sub-division: ....
3. Name of the Block: ....
4. Name of the RPC/village: ....
5. Total number of households in the village: ....
6. Number of households rehabilitated though rubber plantation scheme in the village: ....

Section- A

A. Human capital

1. Total members of the Household: ....

Male (M) Female (F)	Age	Education Qualification	Occupation /major livelihood	Any other	**Gender:** 1. Male (**M**)
					2. Female (**F**)
					Education:
					1. illiterate
					2. nursery
					3. primary (I to iv)
					4. elementary (v to vii)
					5. secondary (viii to x)
					6. higher secondary (xi to xii)
					7. graduate
					8. post graduate
					9. other mention it
					Major livelihoods
					1. Tapper
					2. Latex collector
					3. Processer of latex
					4. Both 1 & 2
					5. Wage earner
					6. employed (pvt/govt)
					7. other mention it
					Business:
					1. Shop
					2. Vendor
					3. Driver
					4. other mention it

2. **Religion:** 1. Hinduism, 2. Christianity, 3. Islam, 4. Others...................................
3. **Community/Tribe:** 1. Tripura, 2. Reang, 3. Debbarma, 4. Jamatia, 5. any other...
4. **House type:** 1.Kaccha (mud wall with tin roof), 2.Half wall with bamboo wall with tin roof, 3. Brick (pucca) wall with tin roof, 4. RCC (pucca), 5. Others..

5. **Type of Sanitation:** 1. Pucca, 2. Mud wall, 3. Bamboo, 4. Open, 5. Piped sewer system, 6. Septic tank, 7. Pit, 8. Public, 9. Others..
6. **What modern technologies and training connected with rubber are you having access to:** ....
7. **What other modern technique or skills do you or any members of the households know/practices:** ....
8. **Do you** or **any members of the household still know/ practice traditional Skills?:** 1. Yes................ 2. No.............. if **Yes** like: 1. Handicraft, 2. Weaving, 3. Pottery, 4. Folk artist, 5. Priest, 6. Shifting cultivation, 7. Fishing & hunting and 6. others.......................
9. Rearing of livestock?: 1. Yes.............. 2. No .... if Yes like: 1. Rearing hen, 2. Rearing pig, 3. Rearing cattle, 4. Rearing goat, 5. Rearing duck, 6. other......................
10. **What are the main sources of livelihood/income of you or the household in the past:** ....

B. Financial capital

1. **Total Monthly/Annual income of the household:** Rs..
2. **Monthly /Annual income from (if) working in rubber plantation:** Rs.............
4. **Monthly/Annual income, if employed in rubber processing centre:** Rs...........
5. **Monthly/Annual income from non-farm activities like:** 1. **Shop** Rs..............., 2. **Driver** Rs....................,3.**Decorator** Rs.................. 4. **Mechanic** Rs.................., 5. **Carpentry** Rs.................., 6. **Petty business in local market** Rs.......................

B.1. Expenditure

1. **Monthly/Annual expenditure in education of children: Rs**.............................
2. **Monthly/Annual household expenditures on food items:** Rs..........................
3. **Annual Expenditure on medical treatment:** Rs..
4. **Annual Expenditure on others household articles: Rs**..

C. Natural capital

1. **Total land holding size of the household in kani/hectare:** ....
2. **Nature of land ownership:** 1.Owner, 2.Moratage to the Corporation, and 3.any other ....
3. **Land under:** 1. Kitchen garden, 3. Wet paddy, 4. Pond, 5. any other.................
4. **Access to land for shifting cultivation:** 1. Yes......... 2. No...........
5. **Access to forest area for collection of household items/needs:** 1. Yes......... 2. No............

6. **Access to stream and water bodies: 1.** Yes......... **2.** No...........
7. **Do you/household have fish pond for growing fishery: 1.** Yes......... **2.** No.........
8. **Has the value of land increased/decreased with the introduction of rubber plantation: 1.** Yes.......... **2.**No............

D. Physical capital

1. **Access to road:** 1. blacktop road, 2. brick road, and 3.mud/kucha road
2. **Access to infrastructures facilities like:** 1. rubber processing & storage centre, 2. training and capacity building, 3. rubber board office, 4. All of these.....................
3. **Access to any Research and development infrastructure facilities:** 1. Yes......... 2. No..........
4. **Access to market:** 1.within 5 Km, 2.More than 5 Km
5. **Access to school:** 1. Yes............ **2.** No............
6. **Access of electricity in the house:** 1. Yes............ **2.** No.............
7. **Access to safe drinking water source:** 1. Hand pump, 2. Water supply, 3. pumped well, 4.well, 5.any others ....
8. **Safe drinking water asset within the House:** 1.Yes.................2.No................

if **Yes** like i. water filter, ii-any other..

9. **Animals assets owned:** 1. Hens, 2.goats, 3.pigs, 4.ducks, 5.cattle, 6-any other......
10. **Mention other assets owned by you/the household:** ....

E. Social capital

1. **Relationship with the neighbor families:** ....
2. **Relationship with the officials of the TRPCL:** ....
3. **Access to (are you part of):** 1. Rubber Grower Societies (RGS), 2. Rubber Producer Societies (RPS), 3. Beneficiary committee member, 4. rubber marketing societies, 5.Self help groups, 5. Any other...
4. **Do you receive any help and facilities from the corporation/agency:** 1. Yes....... 2. No.........
5. **Access to any local development institutions provided by the govt.** 1. Yes......... 2. No...........
6. **Access to training and capacity development activities:** 1. Yes.......... 2.No........ if Yes like 1. Research & Development facilities, 2. training in tapping, 3. training in processing, 4. Any other...
7. **Part of committee members in the locality:** 1. Yes........... 2. No. ....
8. **Are there any (access to) groups/organizations/networks in the local area other the rubber growing agencies:** 1.Yes...2.No...if **Yes** mention..................................

9. **Participation in programme or meeting organized by rubber growing agencies and others government and non-government organization regarding rubber and others:** 1.Yes.......... 2.No............

10. **Apart from rubber plantation development scheme, do any other government schemes or policies (like MGNREGA, Indira Awas Yogana, National rural livelihood mission, Prime Minister's Employment Generation Programme, SWABALAMAN Programme of Government of Tripura etc) implemented in the area to (strengthening or undermining) the livelihoods of the local communities:** 1.Yes...........2.No...........If **Yes** mention ....

Section- B

Group discussion

1. **Is there any conflicts/disputes between the rubber growing agencies and the local people regarding rubber cultivation in the area:** 1.Yes........... 2.No.............

if Yes mention the reason ....

2. **If you/any member of the household work in rubber processing centre, do protective measure used: 1.**Yes............... **2.**No...............
3. **If you/any member of the household work in rubber plantation, during latex collection do any protective measure is put on: 1.**Yes.......... **2.**No.............
4. **Main source of earning before rubber was introduced like:** 1. wage labour,

2. *jhum* cultivation, 3. Any other..

5. **If you/any member of the household work in rubber plantation and rubber processing centre what was the monthly/annual income of the household before rubber was introduced:** Rs..
6. **Attitude and habits like:** 1. Vulnerability/risk of being harm, 2. Misguides by anti-socials, 3.Any other ....
7. **How was the standard of education before introduction of rubber plantation?:** 1. Good.......... 2. Poor......... 3. large scale illiteracy..
8. **How was the infrastructure facilities of the area after introduction of rubber plantation: 1.** Good................**2.** Poor.................**3.** Very poor/bad..................
   About rubber plantation development scheme

1. **What improvement has the rubber plantation development scheme brought to the household, the community and the area:**
2. **How is the area now after the implementation of rubber plantation?:**
3. **Any problem associated with rubber plantation development scheme: 1.** Yes.......... **2.** No...if **Yes** what changes and priorities and relevant initiatives can be brought in this rubber plantation development scheme in the area:
4. **What services and activities would you like the rubber corporation and the state government to provide for present and future betterment of the household, the community and the area:**
5. **What were your view and opinion about rubber plantation and its scheme in the beginning/before and what do you view and feel now about rubber plantation?:**

Color Plate

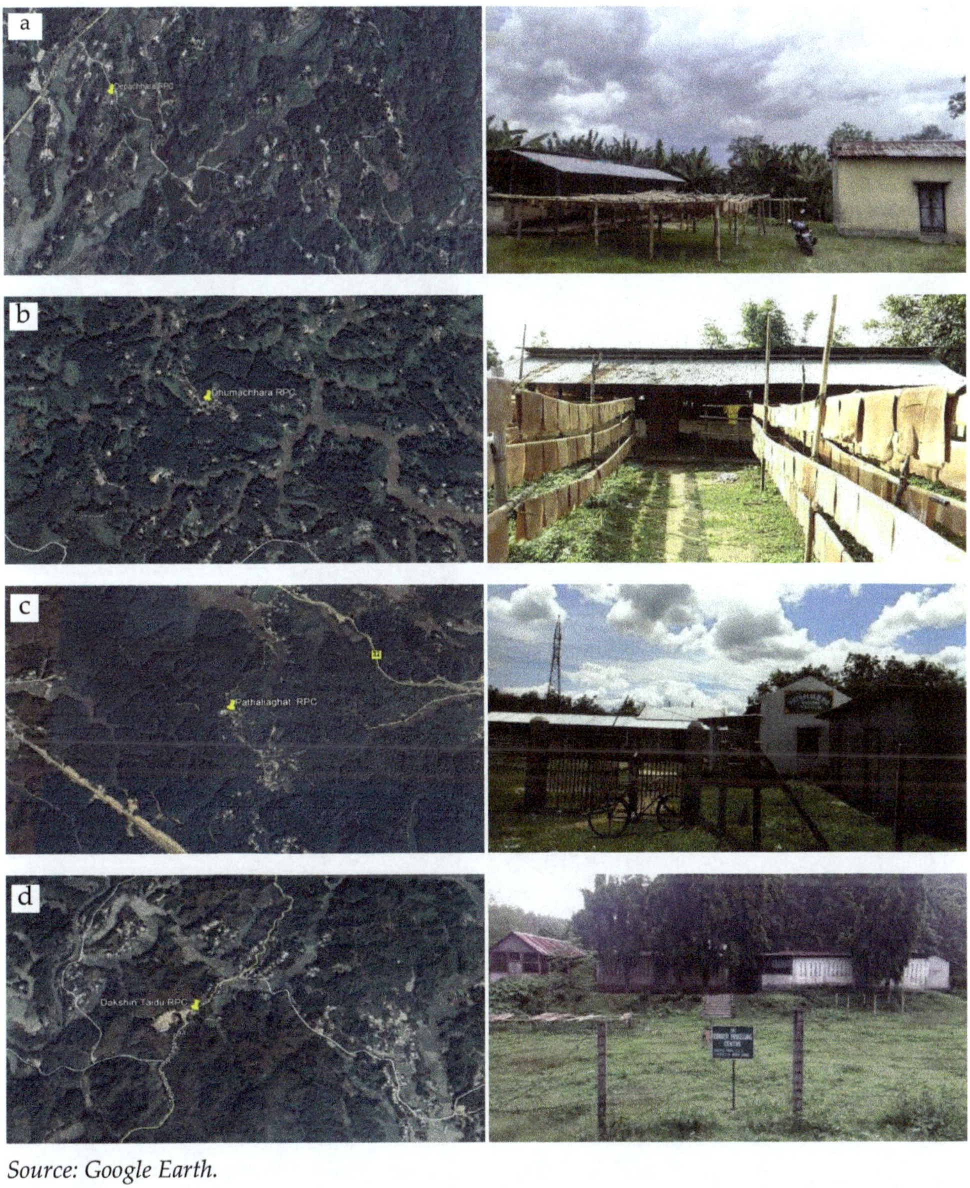

Source: Google Earth.

Plate 1: Image of the surveyed RPCs/villages: (a) Depachhara RPC, (b) Dakshin Dhumachhara RPC, (c) Pathaliaghat RPC, and (d) Dakshin Taidu RPC).

Plate 2: Informal interviews, household surveys: (a), (b), (c) and (d).

Plate 3: Household amenities and possession of consumer items: (a), (b), (c) and (d) shows different types of houses, (e) Access to handpump, (f) Access to Television, (g) Motorcycle owned, and (h) Firewood used as fuel for cooking.

Plate 4: Stages of rubber processing: (a) Latex, (b) Roller machine for processing latex to rubber sheets, (c) rubber sheets, (d) drying of rubber sheets, (e) smoking of rubber sheets, (f) and (g) smoked rubber sheets, and (h) packed rubber sheets for selling/marketing.

Plate 5: Shops and livestocks: (a) and (b) Shops, (c) Mechanic shop, (d) Rearing pig, (e) Rearing ducks, (f) Rearing cow, (g) Rearing goat, and (h) Rearing hens.

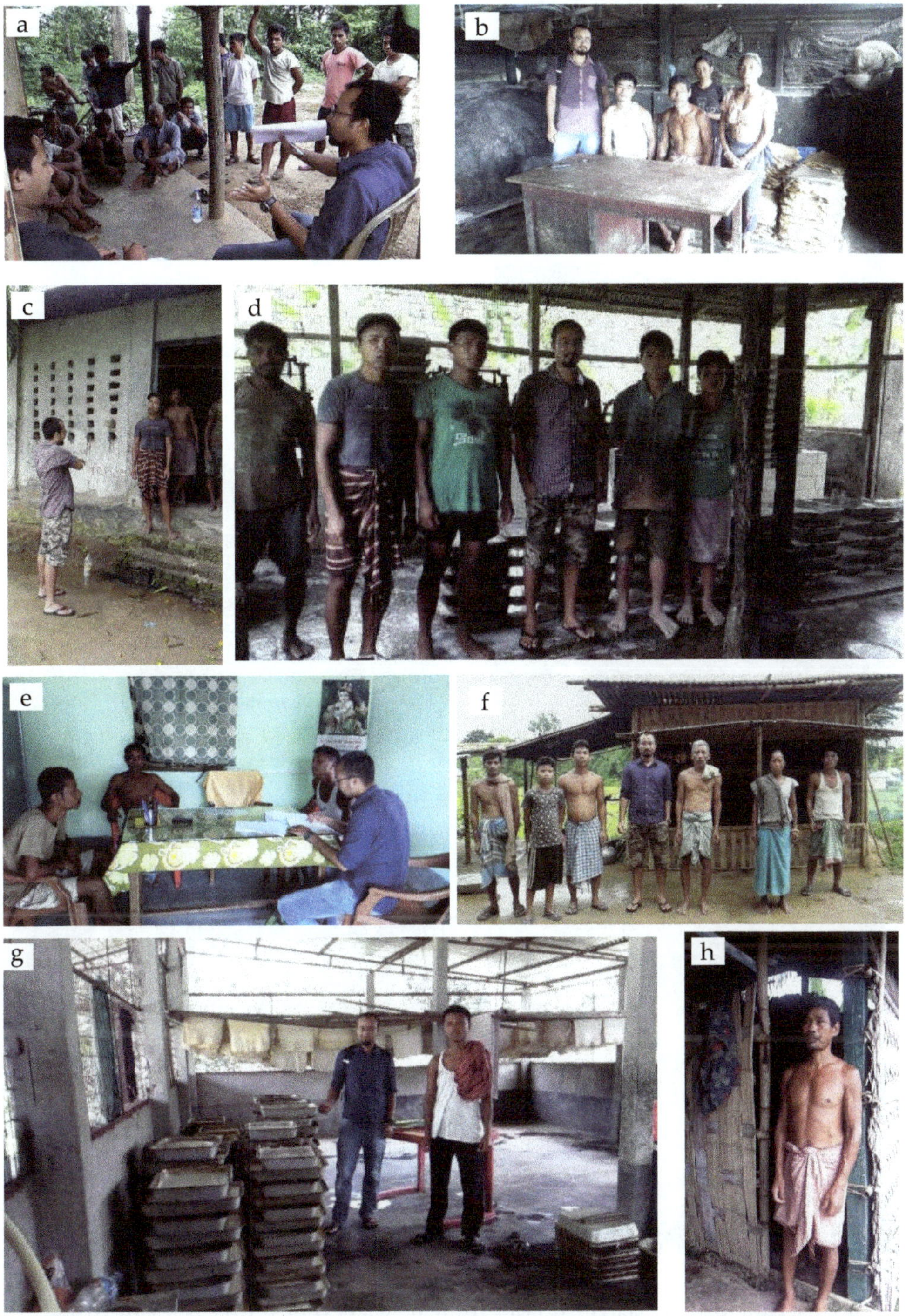

Plate 6: Group discussion: at (a) Dakshin Dhumachhara RPC, (b) Pathaliaghat RPC,(c) Depachhara RPC, and (d) Dakshin Taidu RPC, (e), (f) and (g) discussion with NBHHs members working in rubber processing centres, and (h) member of NBHHs employed as guard in rubber processing centre in Depachhara RPC.

Plate 7: Access to assets: (a) Weaving, (b) Rubber plantation, (c) Pond, (d) Stream, (e) Blacktop road, (f) Market, (g) Safe drinking water source, (h) Electricity, (i) and (j) Schools.

www.ingramcontent.com/pod-product-compliance
Ingram Content Group UK Ltd.
Pitfield, Milton Keynes, MK11 3LW, UK
UKHW021322190726
13845UKWH00003B/23